VOLUME	EDITOR-IN-CHIEF	PAGES	
Collective Vol. IV A revised edition of Annual Volumes 30-39			
	NORMAN RABJOHN, *Editor-in-Chief*	1036	
40	MELVIN S. NEWMAN	114	*Out of print*
41	JOHN D. ROBERTS	118	*Out of print*
42	VIRGIL BOEKELEHIDE	118	*Out of print*
43	B. C. MCKUSICK	124	*Out of print*
44	THE LATE WILLIAM E. PARHAM	131	*Out of print*
45	WILLIAM G. DAUBEN	118	*Out of print*
46	E. J. COREY	146	*Out of print*
47	WILLIAM D. EMMONS	140	*Out of print*
48	PETER YATES	164	*Out of print*
49	KENNETH B. WIBERG	124	*Out of print*
Collective Vol. V A revised edition of Annual Volumes 40-49			
	HENRY E. BAUMGARTEN, *Editor-in-Chief*		
Cumulative Indices to Collective Volumes, I, II, III, IV, V			
	RALPH L. AND THE LATE RACHEL H. SHRINER, .		
50	RONALD BRESLOW	136	*Out of print*
51	RICHARD E. BENSON	209	*Out of print*
52	HERBERT O. HOUSE	192	*Out of print*
53	ARNOLD BROSSI	193	*Out of print*
54	ROBERT E. IRELAND	155	*Out of print*
55	SATORU MASAMUNE	150	*Out of print*
56	GEORGE H. BÜCHI	144	*Out of print*
57	CARL R. JOHNSON	135	*Out of print*
58	THE LATE WILLIAM A. SHEPPARD	216	*Out of print*
59	ROBERT M. COATES	267	*Out of print*
Collective Vol. VI A revised edition of Annual Volumes 50-59			
	WAYLAND E. NOLAND, *Editor-in Chief*	1208	
60	ORVILLE L. CHAPMAN	140	*Out of print*
61	THE LATE ROBERT V. STEVENS	165	*Out of print*
62	MARTIN F. SEMMELHACK	269	*Out of print*
63	GABRIEL SAUCY	291	*Out of print*
64	ANDREW S. KENDE	308	
Collective Vol VII A revised edition of Annual Volumes 60-64			
	JEREMIAH P. FREEMAN, *Editor-in-Chief*	602	
65	EDWIN VEDEJS	278	
66	CLAYTON H. HEATHCOCK	265	
67	BRUCE E. SMART	289	
68	JAMES D. WHITE	318	
69	LEO A. PAQUETTE	328	
70	ALBERT I. MEYERS	305	

Collective Volumes, Cumulative Indices, Annual Volumes 65-70, and Reaction Guide are available from John Wiley & Sons, Inc.

ORGANIC SYNTHESES

ORGANIC SYNTHESES

AN ANNUAL PUBLICATION OF SATISFACTORY
METHODS FOR THE PREPARATION
OF ORGANIC CHEMICALS

VOLUME 70
1992

BOARD OF EDITORS

ALBERT I. MEYERS, *Editor-in-Chief*

ROBERT K. BOECKMAN, Jr. AMOS B. SMITH III
DAVID L. COFFEN JAMES D. WHITE
LARRY E. OVERMAN EKKEHARD WINTERFELDT
LEO A. PAQUETTE HISASHI YAMAMOTO
ICHIRO SHINKAI

THEODORA W. GREENE, *Assistant Editor*
JEREMIAH P. FREEMAN, *Secretary to the Board*
DEPARTMENT OF CHEMISTRY, University of Notre Dame,
Notre Dame, Indiana 46556

ADVISORY BOARD

RICHARD T. ARNOLD ALBERT ESCHENMOSER WAYLAND E. NOLAND
HENRY E. BAUMGARTEN IAN FLEMING RYOJI NOYORI
RICHARD E. BENSON CLAYTON H. HEATHCOCK CHARLES C. PRICE
VIRGIL BOEKELHEIDE E. C. HORNING NORMAN RABJOHN
RONALD BRESLOW HERBERT O. HOUSE JOHN D. ROBERTS
ARNOLD BROSSI ROBERT E. IRELAND GABRIEL SAUCY
GEORGE H. BÜCHI CARL R. JOHNSON DIETER SEEBACH
T. J. CAIRNS WILLIAM S. JOHNSON MARTIN F. SEMMELHACK
JAMES CASON ANDREW S. KENDE RALPH L. SHRINER
ORVILLE L. CHAPMAN N. J. LEONARD BRUCE E. SMART
ROBERT M. COATES B. C. MCKUSICK H. R. SNYDER
E. J. COREY SATORU MASAMUNE EDWIN VEDEJS
WILLIAM G. DAUBEN WATARU NAGATA KENNETH B. WIBERG
WILLIAM D. EMMONS MELVIN S. NEWMAN PETER YATES

FORMER MEMBERS OF THE BOARD, NOW DECEASED

ROGER ADAMS NATHAN L. DRAKE C. R. NOLLER
HOMER ADKINS L. F. FIESER W. E. PARHAM
C. F. H. ALLEN R. C. FUSON R. S. SCHREIBER
WERNER E. BACHMANN HENRY GILMAN JOHN C. SHEEHAN
A. H. BLATT CLIFF S. HAMILTON WILLIAM A. SHEPPARD
WALLACE H. CAROTHERS W. W. HARTMAN LEE IRVIN SMITH
H. T. CLARKE JOHN R. JOHNSON ROBERT V. STEVENS
J. B. CONANT OLIVER KAMM MAX TISHLER
ARTHUR C. COPE C. S. MARVEL FRANK C. WHITMORE

JOHN WILEY & SONS, INC.

NEW YORK • CHICHESTER • BRISBANE • TORONTO • SINGAPORE

Published by John Wiley & Sons, Inc.

Copyright © 1992 by Organic Syntheses, Inc.

All rights reserved. Published simultaneously in Canada.

Reproduction or transiation of any part of this work beyond that permitted by Section 107 or 108 of the 1976 United States Copyright Act without the permission of the copyright owner is unlawful.

"John Wiley & Sons, Inc. is pleased to publish this volume of Organic Syntheses on behalf of Organic Syntheses, Inc. Although Organic Syntheses, Inc. has assured us that each preparation contained in this volume has been checked in an independent laboratory and that any hazards that were uncovered are clearly set forth in the write-up of each preparation, John Wiley & Sons, Inc. does not warrant the preparations against any safety hazards and assumes no liability with respect to the use of the preparations."

Library of Congress Catalog Card Number: 21-17747
ISBN 0-471-57743-X

Printed in the United States of America

10 9 8 7 6 5 4 3 2

NOTICE

With Volume 62, the Editors of *Organic Syntheses* began a new presentation and distribution policy to shorten the time between submission and appearance of an accepted procedure. The soft cover edition of this volume is produced by a rapid and inexpensive process, and is sent at no charge to members of the Organic Divisions of the American and French Chemical Society, The Perkin Division of the Royal Society of Chemistry, and The Society of Synthetic Organic Chemistry, Japan. The soft cover edition is intended as the personal copy of the owner and is not for library use. A hard cover edition is published by John Wiley and Sons Inc. in the traditional format, and differs in content primarily in the inclusion of an index. The hard cover edition is intended primarily for library collections and is available for purchase through the publisher. Annual Volumes 60-64 have been incorporated into a new five-year version of the collective volumes of *Organic Syntheses* which has appeared as *Collective Volume Seven* in the traditional hard cover format. It is available for purchase from the publishers. The Editors hope that the new *Collective Volume* series, appearing twice as frequently as the previous decennial volumes, will provide a permanent and timely edition of the procedures for personal and institutional libraries. The Editors welcome comments and suggestions from users concerning the new editions.

NOMENCLATURE

Both common and systematic names of compounds are used throughout this volume, depending on which the Editor-in-Chief felt was more appropriate. The *Chemical Abstracts* indexing name for each title compound, if it differs from the title name, is given as a subtitle. Systematic *Chemical Abstracts* nomenclature, used in both the 9th and 10th Collective Indexes for the title compound and a selection of other compounds mentioned in the procedure, is provided in an appendix at the end of each preparation. Registry numbers, which are useful in computer searching and identification, are also provided in these appendixes. Whenever two names are concurrently in use and one name is the correct *Chemical Abstracts* name, that name is preferred.

SUBMISSION OF PREPARATIONS

Organic Syntheses welcomes and encourages submission of experimental procedures which lead to compounds of wide interest or which illustrate important new developments in methodology. The Editorial Board will consider proposals in outline format as shown below, and will request full experimental details for those proposals which are of sufficient interest. Submissions which are longer than three steps from commercial sources or from existing *Organic Syntheses* procedures will be accepted only in unusual circumstances.

Organic Syntheses Proposal Format
1) Authors
2) Title
3) Literature reference or enclose preprint if available
4) Proposed sequence
5) Best current alternative(s)
6) a. Proposed scale, final product:
   b. Overall yield:
   c. Method of isolation and purification:
   d. Purity of product (%):
   e. How determined?
7) Any unusual apparatus or experimental technique:

8) Any hazards?
9) Source of starting material?
10) Utility of method of usefulness of product.

Submit to: Dr. Jeremiah P. Freeman, Secretary
            Department of Chemistry
            University of Notre Dame
            Notre Dame, IN 46556

Proposals will be evaluated in outline form, again after submission of full experimental details and discussion, and, finally by checking experimental procedures. A form that details the preparation of a complete procedure (Notice to Submitters) may be obtained from the Secretary.

Additions, corrections, and improvements to the preparations previously published are welcomed; these should be directed to the Secretary. However, checking of such improvements will only be undertaken when new methodology is involved. Substantially improved procedures have been included in the Collective Volumes in place of a previously published procedure.

ACKNOWLEDGMENT

Organic Syntheses wishes to acknowledge the contributions of Hoffmann-La Roche, Inc. and Merck & Co. to the success of this enterprise through their support, in the form of time and expenses, of members of the Boards of Directors and Editors.

DISPOSAL OF CHEMICAL WASTE

General Reference: *Prudent Practices for Disposal of Chemicals from Laboratories*, National Academy Press, Washington, D.C. 1983

Effluents from synthetic organic chemistry fall into the following categories:

1. Gases

1a. Gaseous materials either used or generated in an organic reaction.
1b. Solvent vapors generated in reactions swept with an inert gas and during solvent stripping operations.
1c. Vapors from volatile reagents, intermediates and products.

2. Liquids

2a. Waste solvents and solvent solutions of organic solids (see item 3b).
2b. Aqueous layers from reaction work-up containing volatile organic solvents.
2c. Aqueous waste containing non-volatile organic materials.
2d. Aqueous waste containing inorganic materials.

3. Solids

3a. Metal salts and other inorganic materials.
3b. Organic residues (tars) and other unwanted organic materials.
3c. Used silica gel, charcoal, filter aids, spent catalysts and the like.

The operation of industrial scale synthetic organic chemistry in an environmentally acceptable manner* requires that all these effluent categories be dealt with properly. In small scale operations in a research or academic setting, provision should be made for dealing with the more environmentally offensive categories.

---

*An environmentally acceptable manner may be defined as being both in compliance with all relevant state and federal environmental regulations *and* in accord with the common sense and good judgement of an environmentally aware professional.

1a. Gaseous materials that are toxic or noxious, e.g., halogens, hydrogen halides, hydrogen sulfide, ammonia, hydrogen cyanide, phosphine, nitrogen oxides, metal carbonyls, and the like.
1b. Vapors from noxious volatile organic compounds, e.g., mercaptans, sulfides, volatile amines, acrolein, acrylates, and the like.
2a. All waste solvents and solvent solutions of organic waste.
2c. Aqueous waste containing dissolved organic material known to be toxic.
2d. Aqueous waste containing dissolved inorganic material known to be toxic, particularly compounds of metals such as arsenic, beryllium, chromium, lead, manganese, mercury, nickel, and selenium.
3. All types of solid chemical waste.

Statutory procedures for waste and effluent management take precedence over any other methods. However, for operations in which compliance with statutory regulations is exempt or inapplicable because of scale or other circumstances, the following suggestions may be helpful.

Gases:

Noxious gases and vapors from volatile compounds are best dealt with at the point of generation by "scrubbing" the effluent gas. The gas being swept from a reaction set-up is led through tubing to a (large!) trap to prevent suckback and on into a sintered glass gas dispersion tube immersed in the scrubbing fluid. A bleach container can be conveniently used as a vessel for the scrubbing fluid. The nature of the effluent determines which of four common fluids should be used: dilute sulfuric acid, dilute alkali or sodium carbonate solution, laundry bleach when an oxidizing scrubber is needed, and sodium thiosulfate solution or diluted alkaline sodium borohydride when a reducing scrubber is needed. Ice should be added if an exotherm is anticipated.

Larger scale operations may require the use of a pH meter or starch/iodide test paper to ensure that the scrubbing capacity is not being exceeded.

When the operation is complete, the contents of the scrubber can be poured down the laboratory sink with a large excess (10-100 volumes) of water. If the solution is a large volume of dilute acid or base, it should be neutralized before being poured down the sink.

Liquids:

Every laboratory should be equipped with a waste solvent container in which *all* waste organic solvents and solutions are collected. The contents of these containers should be periodically transferred to properly labeled waste solvent drums and arrangements made for contracted disposal in a regulated and licensed incineration facility.**

Aqueous waste containing dissolved toxic organic material should be decomposed *in situ*, when feasible, by adding acid, base, oxidant, or reductant. Otherwise, the material should be concentrated to a minimum volume and added to the contents of a waste solvent drum.

Aqueous waste containing dissolved toxic inorganic materials should be evaporated to dryness and the residue handled as a solid chemical waste.

Solids:

Soluble organic solid waste can usually be transferred into a waste solvent drum, provided near-term incineration of the contents is assured.

Inorganic solid wastes, particularly those containing toxic metal compounds, used Raney nickel, manganese dioxide, etc. should be placed in glass bottles or lined fiber drums, sealed, properly labeled, and arrangements made for disposal in a secure landfill.** Used mercury is particularly pernicious and small amounts should first be amalgamated with zinc or combined with excess sulfur to solidify the material.

Other types of solid laboratory waste including used silica gel and charcoal should also be packed, labeled, and sent for disposal in a secure landfill.

Special Note:

Since local ordinances may vary widely from one locale to another, one should always check with appropriate authorities. Also, professional disposal services differ in their requirements for segregating and packaging waste.

**If arrangements for incineration of waste solvent and disposal of solid chemical waste by licensed contract disposal services are not in place, a list of providers of such services should be available from a state or local office of environmental protection.

PREFACE

In keeping with tradition, Volume 70 contains 31 procedures, carefully checked by independent laboratories, dealing with important compounds of varied nature. For the first time, this volume has added a new section to each procedure—"Waste Disposal Information." This includes any special disposal of chemical waste utilized by the submitters in order to comply with modern, clean disposal practices of toxic materials. If no unusual disposal method was employed, the editor has inserted the following statement, "All toxic materials were disposed of in accordance with *Prudent Practices for Disposal of Chemicals from Laboratories*, National Academy Press; Washington, DC, 1983." This monograph should be consulted by all users of *Organic Syntheses* procedures in keeping with current trends to preserve the environment. A synopsis of the disposal guidelines are included just prior to this preface.

Since protein and peptide syntheses are rapidly falling into the hands of organic chemists, four procedures to reach building blocks are presented first. N-protected α-amino acids in the form of its β-lactone; **Nα-(BENZYLOXYCARBONYL)-β-(PYRAZOL-1-YL)-L-ALANINE** and the preparation of both β-lactones, **N-tert-BUTOXYCARBONYL-L-SERINE-β-LACTONE** and the **p-TOLUENESULFONIC ACID SALT OF (S)-3-AMINO-2-OXETANONE** are described starting from **N-tert-BUTOXYCARBONYL-L-SERINE**. A useful S or R serine derivative, **1,1-DIMETHYLETHYL (S) or (R)-4-FORMYL-2,2-DIMETHYL-3-OXAZOLIDINECARBOXYLATE** is described followed by the improved preparation of **N-(BENZYLOXYCARBONYL)-L-VINYLGLYCINE METHYL ESTER**, an important intermediate as well as an important substance in biological processes.

The continued fascination chemists possess with asymmetric synthesis provides the basis for the next four procedures. The synthesis of **(R)-(-)-10-METHYL-1(9)-OCTALONE-2** is a nice demonstration of an asymmetric Michael addition by a chiral imine followed by an aldol—in short an asymmetric Robinson annulation. The asymmetric glycolization to **STILBENE DIOL (R,R-1,2-DIPHENYL-1,2-ETHANEDIOL)** represents an olefin oxidation using catalytic alkaloids in tandem with osmium tetroxide. As reagents for a variety of asymmetric alkylations, the preparation of **2-CYANO-6-PHENYLOXAZOLOPIPERIDINE** is presented as well as another route to

xiii

(S)- AND (R)-1,1'-BI-2-NAPHTHOL, obtained by resolution of its pentanoate ester with cholesterol esterase.

In accordance with the editorial board's recent policy reflecting its interest in obtaining more procedures for important heterocyclic compounds, seven procedures for heterocyclic ring syntheses and three of substituent modification follow: **3,4-DIETHYLPYRROLE** and the highly symmetric **OCTAETHYLPORPHYRIN** are described making this synthetic porphyrin readily available for biological modeling and inorganic chemical applications.

This is followed by an example of a pyrrole synthesis: **DIMETHYL-3-PHENYLPYRROLE-2,5-DICARBOXYLATE** is obtained via an inverse demand Diels-Alder synthesis using **DIMETHYL-1,2,4,5-TETRAZINE-3,6-DICARBOXYLATE**. A rhodium-catalyzed cyclization of acetylenes with α-diazocarbonyls provides a nice route to **ETHYL 2-METHYL-5-PHENYL-3-FURAN CARBOXYLATE** and related 2,3,5-trisubstituted furans. Two procedures dealing with cyclization to nitrogen heterocycles are included—the first is an example of iodolactamization to give **8-EXO-IODO-2-AZABICYCLO[3.3.0]OCTAN-3-ONE** and the second procedure is a nice demonstration of an alkyne-iminium cyclization furnishing **(E)-1-BENZYL-3-(1-IODOETHYLIDENE)PIPERIDINE**. A sulfur heterocycle, **9-THIABICYCLO[3.3.1]NONANE-2,6-DIONE** is also included which has shown to be a versatile intermediate to a number of other sulfur heterocycles. The current interest in cryptands for complexing metal ions provides the rationale for the procedure describing **(TRIAZA-21-CROWN-7)4-BENZYL-10,19-DIETHYL-4,10,19-TRIAZA-1,7,13,16-TETRAOXACYCLOHENEICOSANE**.

Modification and elaboration of heterocycles are demonstrated by the transformation of dihydroisoquinolines to **7,8-DIMETHOXY-1,3,4,5-TETRAHYDRO 2H,3-BENZAZEPIN-2-ONE**, the conversion of 2-bromopyrrole to **N-tert-BUTOXY-2-TRIMETHYLSILYLPYRROLE**, and substitution of the α-phenylsulfonyl group in pyran to furnish **TETRAHYDRO-2-(PHENYLETHYNYL)-2H-PYRAN**.

The next group of procedures appearing herein reflects a variety of useful new reagents for performing a variety of synthetic tasks. The preparation of **TRIS(TRIMETHYLSILYL)SILANE** as a hydrogen donor source in free radical chemistry is a welcome addition, as is the preparation of **9-BORABICYCLO[3.3.1]NONANE DIMER**, the well recognized hydroboration agent. A stable methylene transfer agent, **IRON (1$^+$), DICARBONYL(η^5-2,4-CYCLOPENTADIEN-1-YL) (DIMETHYLSULFONIUMη-METHYLIDE)-,TETRAFLUOROBORATE(1$^-$)** is described and shows its utility in the synthesis of **1,1-DIPHENYLCYCLOPROPANE**. The clean

1,2-addition of **RCu(CN)ZnI** reagents to α,β-unsaturated aldehydes is nicely demonstrated as well as the 1,4-addition of **ORGANOBIS(CUPRATES)** in a spiroannelation affording **9,9-DIMETHYLSPIRO[4.5]DECAN-7-ONE**. The interesting properties of **ALKYNYL ARYL IODONIUM SALTS** are highlighted in the preparation of the conjugated enyne, **E-5-PHENYL-DO-DEC-5-EN-7-YNE**.

The last group of procedures included in this volume represent efficient means of reaching important and pivotal intermediates for a host of functionalized materials. The simple, single step, synthesis of **2-METHYL-1,3-CYCLOPENTANEDIONE** from cheap materials is presented, followed by a convenient procedure leading to large scale preparation of the functionalized diene, **(E,Z)-1-METHOXY-2-METHYL-3-(TRIMETHYLSILYLOXY)-1,3-PENTADIENE**. Another useful diene, generated under nickel catalysis between a 1,3-dithiolane and a Grignard reagent, is **(E,E)-TRIMETHYL(4-PHENYL-1,3-BUTADIENTYL)SILANE**. A rapid entry into fluorine-containing alkynes is the procedure describing the pyrolysis of α-**ACYLMETHYLENEPHOSPHORANES** which produces a good yield of **ETHYL 4,4,4-TRIFLUORO-2-BUTYNOATE**. C-Acylation of an enolate using methyl cyanoformate provides a convenient source of the α-carbomethoxyoctalone, **METHYL (1α,4Aβ,8Aα)-2-OXO-DECAHYDRO-1-NAPHTHOATE** and represents a good example of generating β-keto esters under mild conditions. The nitrone functionality is featured in a procedure which makes it in a single step from secondary amines and **6-METHYL-2,3,4,5-TETRAHYDROPYRIDINE N-OXIDE** is the example described. Finally, the synthesis of phospholes: **1-PHENYL-2,3,4,5-TETRAMETHYLPHOSPHOLE** is described as an example of the versatility of zirconocene chemistry.

In closing, the editor acknowledges the skills, patience, and dedication of all my colleagues on the Editorial Board of *Organic Syntheses* and their students and coworkers who painstakingly checked all the procedures included in this edition. Furthermore, the guidance, calming effect, and talent of Professor Jeremiah Freeman, Secretary to the Board, in demanding and maintaining the highest standards are truly appreciated. The final, clearly discernable, and attractive product before the reader is due entirely to Dr. Theodora W. Greene, Assistant Editor to *Organic Syntheses*, and the Freeman staff at the University of Notre Dame who typed and prepared the final manuscript.

ALBERT I. MEYERS

Fort Collins, Colorado
May 1991

CONTENTS

Sunil V. Pansare, Gregory Huyer, Lee D. Arnold, and John C. Vederas	1	SYNTHESIS OF N-PROTECTED α–AMINO ACIDS FROM N-(BENZYLOXYCARBONYL)-L-SERINE VIA ITS β–LACTONE: Nα-(BENZYLOXYCARBONYL)-β–(PYRAZOL-1-YL)-L-ALANINE
Sunil V. Pansare, Lee D. Arnold, and John C. Vederas	10	SYNTHESIS OF N-tert-BUTOXYCARBONYL-L-SERINE β–LACTONE AND THE p-TOLUENESULFONIC ACID SALT OF (S)-3-AMINO-2-OXETANONE
Philip Garner and Jung Min Park	18	1,1,-DIMETHYLETHYL (S)- OR (R)-4-FORMYL-2,2-DIMETHYL-3-OXAZOLIDINECARBOXYLATE: A USEFUL SERINAL DERIVATIVE
Michael Carrasco, Robert J. Jones, Scott Kamel, H. Rapoport, and Thien Truong	29	N-(BENZYLOXYCARBONYL)-L-VINYLGLYCINE METHYL ESTER
G. Revial and M. Pfau	35	(R)-(-)-10-METHYL-1(9)-OCTAL-2-ONE
Blaine H. McKee, Declan G. Gilheany, and K. Barry Sharpless	47	(R,R)-1,2-DIPHENYL-1,2-ETHANEDIOL (STILBENE DIOL)
Martine Bonin, David S. Grierson, Jacques Royer, and Henri-Philippe Husson	54	A STABLE CHIRAL 1,4-DIHYDROPYRIDINE EQUIVALENT FOR THE ASYMMETRIC SYNTHESIS OF SUBSTITUTED PIPERIDINES: 2-CYANO-6-PHENYLOXAZOLOPIPERIDINE
Romas J. Kazlauskas	60	(S)-(-)- AND (R)-(+)-1,1'-BI-2-NAPHTHOL
Jonathan L. Sessler, Azadeh Mozaffari, and Martin R. Johnson	68	3,4-DIETHYLPYRROLE AND 2,3,7,8,12,13,17,18-OCTAETHYLPORPHYRIN
Dale L. Boger, James S. Panek, and Mona Patel	79	PREPARATION AND DIELS-ALDER REACTION OF A REACTIVE, ELECTRON-DEFICIENT HETEROCYCLIC AZADIENE: DIMETHYL 1,2,4,5-TETRAZINE-3,6-DICARBOXYLATE. 1,2-DIAZINE AND PYRROLE INTRODUCTION

Authors	Page	Title
Huw M. L. Davies, William R. Cantrell, Jr., Karen R. Romines, and Jonathan S. Baum	93	SYNTHESIS OF FURANS VIA RHODIUM(II) ACETATE-CATALYZED REACTION OF ACETYLENES WITH α–DIAZOCARBONYLS: ETHYL 2-METHYL-5-PHENYL-3-FURAN-CARBOXYLATE
Spencer Knapp and Frank S. Gibson	101	IODOLACTAMIZATION: 8-exo-IODO-2-AZABICYCLO[3.3.0]OCTAN-3-ONE
H. Arnold, L. E. Overman, M. J. Sharp and M. C. Witschel	111	(E)-1-BENZYL-3-(1-IODOETHYLIDENE)-PIPERIDINE: NUCLEOPHILE-PROMOTED ALKYNE-IMINIUM ION CYCLIZATIONS
Roger Bishop	120	9-THIABICYCLO[3.3.1]NONANE-2,6-DIONE
Krzysztof E. Krakowiak and Jerald S. Bradshaw	129	4-BENZYL-10,19-DIETHYL-4,10,19-TRIAZA-1,7,13,16-TETRAOXACYCLO-HENEICOSANE (TRIAZA-21-CROWN-7)
George R. Lenz and Ralph A. Lessor	139	TETRAHYDRO-3-BENZAZEPIN-2-ONES: LEAD TETRAACETATE OXIDATION OF ISOQUINOLINE ENAMIDES
Wha Chen, E. Kyle Stephenson, Michael P. Cava, and Yvette A. Jackson	151	2-SUBSTITUTED PYRROLES FROM N-tert-BUTOXYCARBONYL-2-BROMOPYRROLE: N-tert-BUTOXY-2-TRIMETHYLSILYLPYRROLE
Dearg S. Brown and Steven V. Ley	157	SUBSTITUTION REACTIONS OF 2-BENZENESULFONYL CYCLIC ETHERS: TETRAHYDRO-2-(PHENYLETHYNYL)-2H-PYRAN
Joachim Dickhaut and Bernd Giese	164	TRIS(TRIMETHYLSILYL)SILANE
John A. Soderquist and Alvin Negron	169	9-BORABICYCLO[3.3.1]NONANE DIMER
Matthew N. Mattson, Edward J. O'Connor, and Paul Helquist	177	CYCLOPROPANATION USING AN IRON-CONTAINING METHYLENE TRANSFER REAGENT: 1,1-DIPHENYL-CYCLOPROPANE

Ming Chang P. Yeh, Huai Gu Chen, Paul Knochel	195	1,2-ADDITION OF A FUNCTIONALIZED ZINC-COPPER ORGANOMETALLIC [RCu(CN)ZnI] TO AN α,β–UNSATURATED ALDEHYDE: (E)-2-(4-HYDROXY-6-PHENYL-5-HEXENYL)-1H-ISOINDOLE-1,3(2H)-DIONE
Paul A. Wender, Alan W. White, and Frank E. McDonald	204	SPIROANNELATION VIA ORGANOBIS-(CUPRATES): 9,9-DIMETHYLSPIRO-[4.5]DECAN-7-ONE
Peter J. Stang and Tsugio Kitamura	215	ALKYNYL(PHENYL)IODONIUM TOSYLATES: PREPARATION AND STEREOSPECIFIC COUPLING WITH VINYLCOPPER REAGENTS. FORMATION OF CONJUGATED ENYNES. 1-HEXYNYL(PHENYL)-IODONIUM TOSYLATE AND (E)-5-PHENYLDODEC-5-EN-7-YNE
Philip G. Meister, Matthew R. Sivik, and Leo A. Paquette	226	2-METHYL-1,3-CYCLOPENTANEDIONE
David C. Myles and Mathew H. Bigham	231	PREPARATION OF (E,Z)-1-METHOXY-2-METHYL-3-(TRIMETHYLSILOXY)-1,3-PENTADIENE
Zhi-Jie Ni and Tien-Yau Luh	240	NICKEL-CATALYZED SILYLOLEFIN-ATION OF ALLYLIC DITHIOACETALS: (E,E)-TRIMETHYL(4-PHENYL-1,3-BUTADIENYL)SILANE
B. C. Hamper	246	α-ACETYLENIC ESTERS FROM α–ACYLMETHYLENEPHOSPHORANES: α–ETHYL 4,4,4-TRIFLUOROTETROLATE
Simon R. Crabtree, Lewis N. Mander, and S. Paul Sethi	256	SYNTHESIS OF β–KETO ESTERS BY C-ACYLATION OF PREFORMED ENOLATES WITH METHYL CYANOFORMATE: PREPARATION OF METHYL (1α,4aβ,8aα)-2-OXO-DECAHYDRO-1-NAPHTHOATE
Shun-Ichi Murahashi, Tatsuki Shiota, and Yasushi Imada	265	OXIDATION OF SECONDARY AMINES TO NITRONES: 6-METHYL-2,3,4,5-TETRAHYDROPYRIDINE N-OXIDE
Paul J. Fagan and William A. Nugent	272	1-PHENYL-2,3,4,5-TETRAMETHYL-PHOSPHOLE
Unchecked Procedures	278	

Cumulative Author Index for Volume 70   282

Cumulative Subject Index for Volume 70   284

ORGANIC SYNTHESES

SYNTHESIS OF N-PROTECTED α-AMINO ACIDS FROM N-(BENZYLOXYCARBONYL)-L-SERINE VIA ITS β-LACTONE: Nα-(BENZYLOXYCARBONYL)-β-(PYRAZOL-1-YL)-L-ALANINE

(Serine, N-carboxy-, β-lactone, benzyl ester, L-)

A.

B.

Submitted by Sunil V. Pansare, Gregory Huyer, Lee D. Arnold, and John C. Vederas.[1]

Checked by M. Jones, G. L. Olson, and David L. Coffen.

1. Procedure

A. *N-(Benzyloxycarbonyl)-L-serine β-lactone.*[2] A 2-L, three-necked, round-bottomed flask is equipped with a magnetic stirring bar, an argon inlet adaptor, a low temperature thermometer, and a rubber septum (Note 1). The flask is charged with tetrahydrofuran (1.1 L) and triphenylphosphine (42.1 g, 160 mmol, Note 2). The triphenylphosphine is dissolved with stirring and the flask is then cooled to -78°C with a dry ice-acetone bath (Note 3). Distilled dimethyl azodicarboxylate (17.7 mL, 160 mmol, d = 1.33 g/mL at 25°C) is added dropwise with a syringe over 10 min (*Caution*, Note 4). The resulting pale yellow solution is stirred at -75° to -78°C for 10 min, at which point a milky white slurry is obtained. The rubber septum on the flask is quickly replaced with a 1-L pressure-equalizing dropping funnel containing a solution of N-(benzyloxycarbonyl)-L-serine (38.3 g, 160 mmol) in tetrahydrofuran (240 mL), (Notes 2, 5) which is added dropwise to the mixture over 30 min. After completion of the addition, the mixture is stirred at -75° to -77°C for 20 min, the cooling bath is removed, and the mixture is slowly warmed with stirring to room temperature over 2.5 hr (Note 6). The solvent is removed on a rotary evaporator at 35°C. The residual pale yellow syrup is dried briefly (15 min) under high vacuum (~0.2 mm) and suspended in hexane/ethyl acetate (4/1, 20 mL). Ethyl acetate (30 mL) is added to give a solution which is applied to a 10 x 23-cm column of flash silica gel^3 (800 g) packed in hexane/ethyl acetate (4/1). The flask and the sides of the column are rinsed with additional ethyl acetate (20 mL), and this is added to the column which is then eluted with hexane/ethyl acetate (4/1, 2.6 L). The solvent is changed to hexane/ethyl acetate (3/2) and 500-mL fractions are collected. Concentration of fractions 6-10 on a rotary evaporator gives 15.5 g (44%) of analytically pure N-(benzyloxycarbonyl)-L-serine β-lactone (Note 7). Fractions 11-14 contain slightly impure β-lactone. The solid obtained from concentration of these fractions on a rotary evaporator is dissolved in dichloromethane (50 mL) and precipitated by addition of hexane (50 mL) followed by cooling to -20°C (0.5 hr). The process is repeated twice to afford additional β–lactone (1.25 g). The total yield is 16.8 g (47%) (Note 8).

B. *N^α-(Benzyloxycarbonyl)-β-(pyrazol-1-yl)-L-alanine.*[2,4] A 500-mL, single-necked, round-bottomed flask is equipped with a magnetic stirring bar, a rubber septum and an argon inlet (Note 1). The flask is charged with N-(benzyloxycarbonyl)-L-serine β-lactone (15 g, 68 mmol) and anhydrous acetonitrile (240 mL, Note 2). The cloudy mixture is stirred and solid pyrazole (4.9 g, 72 mmol) (Note 9) is added. The rubber septum is quickly replaced with a reflux condenser and an argon inlet adaptor, and the reaction mixture is heated in an oil bath at 52-54°C for 24 hr. The solvent is removed on a rotary evaporator to leave a white solid that is dried under vacuum for 30 min. Sodium hydroxide (1 N, 69 mL) is added, the suspension is diluted with distilled water (350 mL) and the mixture is stirred vigorously for 5 min. It is then extracted with dichloromethane (3 x 100 mL) to remove unreacted pyrazole and side products. The aqueous phase is cooled in an ice bath to ca. 4°C, and concentrated hydrochloric acid is added with stirring to bring the pH to 1 (about 10 mL required). The resulting precipitate is filtered, washed with water (ca. 75 mL), air dried, and then completely dried in a desiccator over phosphorus oxide (P_2O_5) at 0.2 mm for 12 hr. This material is recrystallized from ethyl acetate (350 mL) to give 7.3 g of pure product (37% yield). Concentration of the mother liquor on a rotary evaporator and recrystallization from the minimum volume of ethyl acetate gives an additional 1.1 g of analytically pure N^α-(benzyloxycarbonyl)-β-(pyrazol-1-yl)-L-alanine. The total yield is 8.4 g (43%) (Note 10).

2. Notes

1. The glass components of the apparatus are dried overnight in a 120°C oven, and then assembled and maintained under an atmosphere of dry argon or nitrogen before use. It is essential to complete the purification of the β–lactone as rapidly as possible because this compound is unstable in the crude reaction mixture.

2. Triphenylphosphine (obtained from General Intermediates of Canada) and N-(benzyloxycarbonyl)-L-serine (obtained from Sigma Chemical Company) were dried under reduced pressure over P_2O_5 for 72 hr and 24 hr, respectively. Acetonitrile was refluxed over calcium hydride (CaH_2) for ca. 10 hr and distilled from CaH_2 before

use. Tetrahydrofuran was distilled from sodium benzophenone ketyl directly into the glassware (under argon) the day before and stored under argon overnight until used.

3. The temperature of the solution should be about -75°C before dimethyl azodicarboxylate is added.

4. Dimethyl azodicarboxylate (manufactured by Tokyo Kasei Kogyo Co., Japan) was purchased from CTC Organic, 792 Windsor Street, Atlanta, GA 30315. Overheating of dimethyl azodicarboxylate should be avoided because of the danger of explosion. Distillation should be conducted from a temperature-controlled bath in the hood behind a safety shield. The material used distilled at 71-72°C (2 mm), at a bath temperature of 84-86°C. It is important that the addition of this compound to the reaction mixture be carried out at a constant rate without interruption because it tends to freeze in the syringe needle. The checkers explored the use of diethyl azodicarboxylate because of its lower cost and wider availability. However, the corresponding hydrazine derivative is more difficult to separate from the β–lactone product of this step.

5. The solution of dried N-(benzyloxycarbonyl)-L-serine was made up separately in an addition funnel under an atmosphere of argon. This avoids complications that may arise if the funnel is prefitted on the reaction vessel.

6. The reaction vessel was placed in a water bath at room temperature after the temperature of the mixture was ca. 15°C.

7. The reaction usually works better on a small scale (25 mmol) and a yield of 60% or more is usually obtained. The flash chromatography column was eluted so that the solvent level dropped 1 cm/13 sec. This corresponds to an approximate rate of 362 mL/min. The concentration and purification of the reaction mixture should be carried out as quickly as possible on the same day. Although storage of the concentrated reaction mixture at -20°C overnight results in substantial decomposition

of the β–lactone, column fractions containing pure β–lactone (after chromatography) can be stored at 4°C overnight and concentrated on the following day.

The β–lactone is readily visualized by TLC (Merck, Kieselgel 60 F_{254}, 0.25-mm thickness, hexane/ethyl acetate (55/45) as solvent system) under UV, or by using bromocresol green spray (0.04% in EtOH, made blue by NaOH) followed by heating of the plate to detect the β–lactone as a yellow spot on a blue background.

8. The product exhibits the following properties: mp 133-134°C; $[\alpha]_D^{22}$ -26.5° (CH_3CN, c 1); IR cm^{-1}: 3355, 1847, 1828, 1685, 1530, 1268; ^1H NMR (360 MHz, CD_2Cl_2) δ: 4.4 (m, 2 H, CH-C$\underline{H}_2$-O), 5.0-5.1 (m, 1 H, N-C$\underline{H}$-CO), 5.12 (s, 2 H, OC$\underline{H}_2$Ph), 5.5-5.7 (br, s, 1 H, N$\underline{H}$), 7.3-7.4 (s, 5 H, Ar$\underline{H}$); EI-MS: M+ 221.0681 (221.0688 calcd for $C_{11}H_{11}NO_4$). Anal. Calcd for $C_{11}H_{11}NO_4$: C, 59.71; H, 4.97, 6.33. Found: C, 59.66; H, 4.92; N, 6.32.

The optical purity was determined as previously described,[2b] and corresponds within experimental error to that of the starting material. N-(Benzyloxycarbonyl)-L-serine obtained from Sigma typically contains 0.75-2.80% of the D-isomer.

9. Pyrazole was obtained from Aldrich Chemical Company, Inc.

10. The reaction usually is more successful on a smaller scale and yields up to 70% can be obtained. The product exhibits the following properties: mp 168-169°C, $[\alpha]_D^{22}$ -53.5°; (DMF, c 1); IR (KBr disc) cm^{-1}: 3350, 1745, 1696, 1534, 1260; ^1H NMR (360 MHz, CD_3OD) δ: 4.4-4.5 (m, 1 H, C$\underline{H}$), 4.60-4.70 (m, 2 H, C$\underline{H}_2$N), 5.08 (s, 2 H, OC$\underline{H}_2$Ph), 6.25 (t, 1 H), 7.3 (s, 5 H, -Ph), 7.48 (d, 1 H), 7.52 (d, 1 H); MS: FAB MS in glycerol m/z 290 (MH+, 36%). Anal. Calcd for $C_{14}H_{15}N_3O_4$: C, 58.13; H, 5.23; N, 14.53. Found: C, 57.86; H, 5.25; N, 14.36.

Waste Disposal Information

All toxic materials were disposed of in accordance with "Prudent Practices for Disposal of Chemicals from Laboratories"; National Academy Press; Washington, DC, 1983.

3. Discussion

A large number of α–amino acids with interesting biological properties occur in nature.[5] This fact, together with the utility of amino acids as chiral synthons, catalysts, and auxiliaries,[6] has stimulated extensive interest in their chemical synthesis.[7] Serine is an especially attractive starting material for preparation of other amino acids because both enantiomers are commercially available (in both free and various N-protected forms) with high optical purity at relatively low cost. Recent work has shown that chiral N-protected serine β–lactones are readily formed under modified Mitsunobu conditions and that they react readily with a variety of carbon, nitrogen, oxygen, sulfur, and halogen nucleophiles to afford optically pure N-protected α–amino acids (Scheme 1).[2,8]

The procedure given here describes the preparation and use of N-(benzyloxycarbonyl)-L-serine β–lactone for the synthesis of a protected β–substituted alanine, N$^\alpha$-(benzyloxycarbonyl)-β--(pyrazol-1-yl)-L-alanine. This compound occurs in watermelon seeds,[9] and has been used as a histidine analog.[4b] Its synthesis illustrates how serine β–lactones can provide convenient access to other β–substituted alanines such as mimosine, willardiine, quisqualic acid, and stizolobic acid which occur in higher plants.[5a,10] Many previous chemical syntheses of racemic pyrazolylalanine have been published; the best of these routes appear to be from acetamidoacrylic acid (94-96% yield)[11] and from O-acetylserine (40-45%).[4c] The racemate has been resolved.[4b] The β–(pyrazol-1-yl)-L-alanine synthase enzyme has been purified and used to make the chiral material from O-acetyl-L-serine.[12]

The crystalline N-(benzyloxycarbonyl)-serine β–lactone is easily handled in air at room temperature and can be stored dry at -20°C for many months without decomposition. Solutions of this compound in non-nucleophilic organic solvents (e.g., chloroform, ethyl acetate) or in neutral or slightly acidic water (pH 3-6) are stable for several days. Aqueous base rapidly hydrolyzes the β-lactone.[2b,8]

Scheme 1

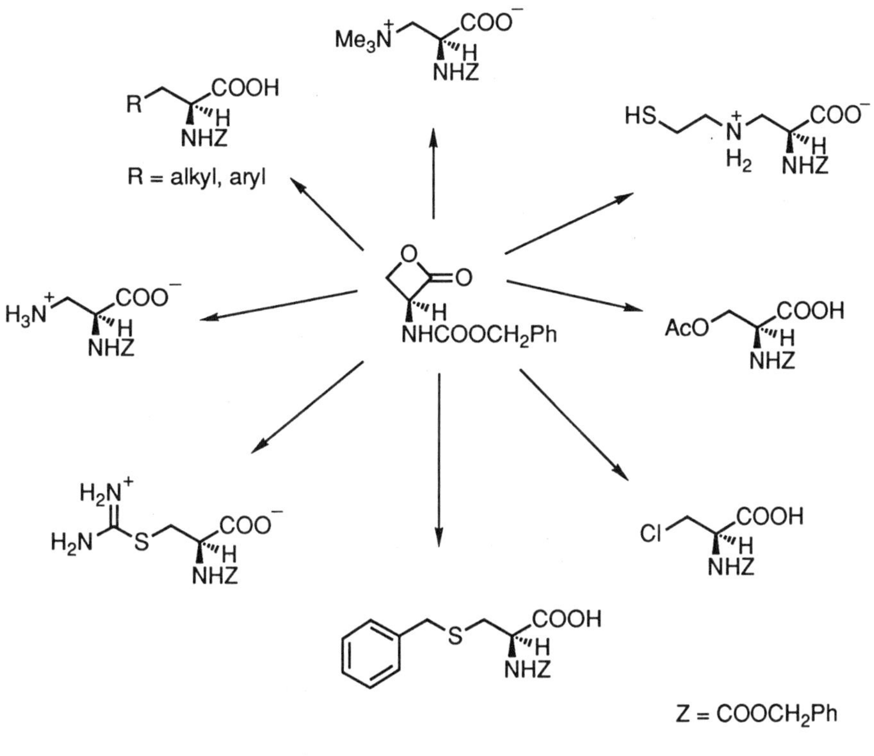

1. Department of Chemistry, University of Alberta, Edmonton, Alberta, Canada, T6G 2G2.
2. (a) Arnold, L. D.; Kalantar, T. H.; Vederas, J. C. *J. Am. Chem. Soc.* **1985**, *107*, 7105; (b) Arnold, L. D.; Drover, J. C. G.; Vederas, J. C. *J. Am. Chem. Soc.* **1987**, *109*, 4649.
3. Still, W. C.; Kahn, M.; Mitra, A. *J. Org. Chem.* **1978**, *43*, 2923.
4. (a) Hofmann, K.; Andreatta, R.; Bohn, H. *J. Am. Chem. Soc.* **1968**, *90*, 6207; (b) Hofmann, K.; Bohn, H. *J. Am. Chem. Soc.* **1966**, *88*, 5914; (c) Murakoshi, I.; Ikegami, F.; Yoneda, Y.; Ihara, H.; Sakata, K.; Koide, C. *Chem. Pharm. Bull.* **1986**, *34*, 1473.
5. (a) "Chemistry and Biochemistry of the Amino Acids"; Barrett, G. C., Ed.; Chapman and Hall: London, 1985; (b) Bender, D. A. "Amino Acid Metabolism", 2nd ed.; Wiley: New York, 1985; (c) "Amino Acids and Peptides"; Davies, J. S., Ed.; Chapman and Hall: London, 1985.
6. (a) Coppola, G. M.; Schuster, H. F. "Asymmetric Synthesis: Construction of Chiral Molecules Using Amino Acids", Wiley: New York, 1987; (b) Martens, J. *Top. Curr. Chem.* **1984**, *125*, 165-246; (c) Ottenheijm, H. C. *Chimia* **1985**, *39*, 89-98.
7. For a large number of leading references see: "α–Amino Acid Synthesis", Tetrahedron Symposia-in-Print Number 33, O'Donnell, M. J.; Ed.; *Tetrahedron* **1988** *44*, 5253-5614.
8. (a) Arnold, L. D.; May, R. G.; Vederas, J. C. *J. Am. Chem. Soc.* **1988**, *110*, 2237-2241; (b) Ramer, S. E.; Moore, R. N.; Vederas, J. C. *Can. J. Chem.* **1986**, *64*, 706-713.
9. Noe, F. F.; Fowden, L. *Biochem. J.* **1960**, *77*, 543-546.
10. Fowden, L.; Lea, P. J.; Bell, E. A. *Adv. Enzymol.* **1979**, *50*, 117-175.
11. Murakoshi, I.; Ohmiya, S.; Haginiwa, J. *Chem. Pharm. Bull* **1972**, *20*, 609-611.

12. Murakoshi, I.; Ikegami, F.; Hinuma, Y.; Hanma, Y. *Phytochemistry* **1984**, *23*, 973-977.

Appendix
Chemical Abstracts Nomenclature (Collective Index Number); (Registry Number)

N-(Benzyloxycarbonyl)-L-serine: L-Serine, N-[(phenylmethoxy)carbonyl]- (9); (1145-80-8)

N-(Benzyloxycarbonyl)-L-serine β–lactone: Serine, N-carboxy-, β–lactone, benzyl ester, L- (9); (26054-60-4)

Triphenylphosphine: Phosphine, triphenyl- (8,9); (603-35-0)

Dimethyl azodicarboxylate: Diazenedicarboxylic acid, dimethyl ester (9); (2446-84-6)

N^{α}-(Benzyloxycarbonyl)-β–(pyrazol-1-yl)-L-alanine: 1H-Pyrazole-1-propanoic acid, α-[[(phenylmethoxy)carbonyl]amino]-, (S)- [New compound: No registry number yet]

Pyrazole (8); 1H-Pyrazole (9); (288-13-1)

SYNTHESIS OF N-tert-BUTOXYCARBONYL-L-SERINE β–LACTONE AND THE p-TOLUENESULFONIC ACID SALT OF (S)-3-AMINO-2-OXETANONE

(Carbamic acid, (2-oxo-3-oxetanyl)- 1,1-dimethylethyl ester, (S)-) and (2-Oxetanone, 3-amino-, (S)-, 4-methylbenzenesulfonate)

Submitted by Sunil V. Pansare, Lee D. Arnold, and John C. Vederas.[1]
Checked by P. S. Manchand, P. Mastrodonato-DeLora, and D. L. Coffen.

1. Procedure

A. *N-(tert-Butoxycarbonyl)-L-serine β–lactone*.[2] A 2-L, three-necked, round-bottomed flask is equipped with a magnetic stirring bar, an argon inlet adaptor, a low temperature thermometer, and a rubber septum (Note 1). The flask is charged with tetrahydrofuran (1.1 L) and triphenylphosphine (42.1 g, 160 mmol) (Note 2). The triphenylphosphine is dissolved with stirring, and the flask is cooled to -78°C with a dry ice-acetone bath maintained at that temperature (Note 3). Distilled diethyl azodicarboxylate (DEAD) (27.86 g, 160 mmol) is then added dropwise with a syringe over 10 min (Note 4). The resulting pale yellow solution is stirred at -75°C to -78°C for 10 min, at which point a milky slurry is obtained. The rubber septum on the flask is quickly replaced with a 1-L pressure equalizing dropping funnel containing a solution of N-(tert-butoxycarbonyl)-L-serine (32.8 g, 160 mmol) in tetrahydrofuran (240 mL), (Notes 2 and 5) which is then added dropwise to the mixture over 30 min. After completion of the addition, the mixture is stirred at -75°C to -78°C for 20 min, the cooling bath is removed, and the mixture is slowly warmed with stirring to room temperature over 2.5 hr (Note 6). The solvent is removed on a rotary evaporator at 35°C. The residual pale yellow syrup is suspended in hexane/ethyl acetate (85/15, 20 mL), slurried, and the solid is removed by filtration. The filtrate is diluted with ethyl acetate (30 mL) to give a solution which is applied to a 10 x 23-cm column of flash silica gel[3] (800 g) packed in hexane/ethyl acetate (85/15). The flask and the sides of the column are rinsed with additional ethyl acetate (20 mL), which is added to the column that is then eluted with hexane/ethyl acetate (85/15, 2.7 L). The solvent is changed to hexane/ethyl acetate (7/3) and 500-mL fractions are collected. Concentration of fractions 6-12 on a rotary evaporator gives 12.07 (40%) of pure N-(tert-butoxycarbonyl)-L-serine β–lactone (Notes 7 and 8).

B. *(S)-3-Amino-2-oxetanone p-toluenesulfonic acid salt.*[4] A 500-mL, single-necked, round-bottomed flask is equipped with a magnetic stirring bar and an argon inlet adaptor (Note 1). The flask is charged with a mixture of N-(tert-butoxycarbonyl)-L-serine β–lactone (14.0 g, 74.8 mmol) and anhydrous p-toluenesulfonic acid (13.5 g, 78.5 mmol) (Note 9). The argon inlet adaptor is replaced with a rubber septum and an argon inlet and the flask is cooled in an ice bath for ca. 15 min. Anhydrous trifluoroacetic acid (200 mL, Note 10) is added by cannula along the sides of the flask over 20 min (stirring is initiated when possible). The pale yellow solution is stirred at

0°C for 10 min, the trifluoroacetic acid is removed on a rotary evaporator at below 30°C, and the resulting syrup is placed under high vacuum (~ 0.2 mm) for ca. 1 hr. Anhydrous ether (200 mL, Note 11) is added to the resulting solid, the mixture is triturated to break up lumps, and the suspension is filtered. The solid thus obtained is washed with ether (100 mL), air dried (5 min), and then dried under reduced pressure (0.2 mm) overnight to give 18.4 g (95% yield) of (S)-3-amino-2-oxetanone p-toluenesulfonic acid salt (Note 12).

2. Notes

1. The glass components of the apparatus are dried overnight in a 120°C oven, and then assembled and maintained under an atmosphere of dry nitrogen or argon before use.

2. Triphenylphosphine (obtained from Aldrich Chemical Company, Inc.) and N-tert-butoxycarbonyl-L-serine (obtained from U.S. Biochemical Corporation) were dried over P_2O_5 for 72 hr and 24 hr, respectively. Tetrahydrofuran was distilled from sodium benzophenone ketyl directly into the glassware (under argon) the day before and stored under argon overnight until used.

3. The temperature of the solution should be about -75°C before the diethyl azodicarboxylate is added.

4. Diethyl azodicarboxylate was obtained from Fluka AG. The material used was distilled, bp 82-83°C at 2 mm, in the hood behind a safety shield. It is important that the addition of this compound to the reaction mixture be done at a constant rate without interruption because of its tendency to freeze in the syringe needle. Dimethyl azodicarboxylate (17.7 mL, 160 mmol; d = 1.33 g/mL at 25°C; bp 71-72°C at 2 mm) can be used in place of diethyl azodicarboxylate. This compound is manfactured by Tokyo Kasei Kogyo Co. and is available from CTC Organics.

5. The solution of dried N-(tert-butoxycarbonyl)-L-serine was made up separately in an addition funnel under an atmosphere of argon. This avoids complications that may arise if the funnel is prefitted on the reaction vessel.

6. The reaction vessel was placed in a water bath at room temperature after the temperature of the mixture was ca. 15°C.

7. The reaction usually works better on a smaller scale (25 mmol) where a yield of 70% or greater can be obtained. The flash column was eluted such that the solvent level dropped 1 cm/13 sec. This corresponds to an approximate rate of 362 mL/min. The concentration and purification of the reaction mixture should be done as quickly as possible on the same day. Although storage of the concentrated reaction mixture at -20°C overnight results in substantial decomposition of the lactone, column fractions containing pure lactone (after chromatography) can be stored at 4°C overnight and concentrated on the following day.

The β–lactone is readily visualized by TLC (Merck, Kieselgel 60 F_{254}, 0.25 mm thickness, hexane/ethyl acetate (65/35) as solvent system) by staining in iodine or by using bromocresol green spray (0.04% in EtOH, made blue by NaOH) followed by heating the plate to detect the lactone as a yellow spot on a blue background.

8. The product exhibits the following properties: mp 119.5-120.5°C; $[\alpha]_D^{24}$ -26.2° (CH_3CN, c 1); IR cm^{-1}: 3358, 1836, 1678, 1533, 1290, 1104; ^{1}H NMR (360 MHz, CD_2Cl_2) δ: 1.45 (s, 9 H, -C(C$\underline{H}_3$)$_3$), 4.4-4.45 (m, 2 H, C$\underline{H}_2$), 4.95-5.05 (m, 1 H, NH-C$\underline{H}$), 5.2-5.4 (br s, 1 H, N$\underline{H}$); MS (CI, NH_3), m/z 205 (M·$NH_4^{\oplus}$, 100%). Anal. Calcd for $C_8H_{13}NO_4$: C, 51.33; H, 6.99; N, 7.48. Found: C, 51.28; H, 7.01; N, 7.42.

9. Anhydrous p-toluenesulfonic acid was prepared from the monohydrate (obtained from Aldrich Chemical Company, Inc.) by solution in hot benzene with the aid of ethyl acetate and azeotropic distillation to 50% volume to remove water. The solution was then concentrated to a syrup on a rotary evaporator. The syrup was dissolved in a minimum volume of acetone and excess benzene was added to

precipitate the anhydrous acid. This material was recrystallized from acetone/benzene and dried at 50°C for 12 hr. It was stored under reduced pressure over anhydrous $CaSO_4$. The material used melted at 104-105°C.

10. Trifluoroacetic acid (obtained from Aldrich Chemical Company, Inc.) was refluxed over P_2O_5 for ca. 3 hr and then distilled from P_2O_5 under an atmosphere of argon. The material used distilled at 68-71°C. All manipulations involving trifluoroacetic acid (except removal on a rotary evaporator) were done in a fume hood.

11. Commercially available anhydrous ether (obtained from Fisher Scientific Company) was used directly from a freshly opened can.

12. The product exhibits the following properties: mp (~ 4°C/min) 133-135°C (darkens), 173°C (dec) (rapid); $[\alpha]_D^{24}$ -15.8° (DMF, c 2.2); IR (Fluorolube mull) cm^{-1}: 3040, 1833, 1547; ^1H NMR (360 MHz, d_7 DMF) δ: 2.3 (s, 3 H, ArC$\underline{H}_3$), 4.66 (m, 1 H, CH$\underline{H}$O), 4.74 (m, 1 H, CH$\underline{H}$O), 5.54 (dd, 1 H, J = 4.6, 6.5, C$\underline{H}$), 7.15 (d, 2 H, J = 8, m-ArH), 7.64-7.7 (d, 2 H, J = 8, o-Ar$\underline{H}$); FAB MS (glycerol) m/z 260 (MH$^\oplus$). Anal. Calcd for $C_{10}H_{13}NO_5S$: C, 46.32; H, 5.05; N, 5.40; S, 12.37. Found: C, 46.44; H, 5.14; N, 5.24; S, 12.41.

Waste Disposal Information

All toxic materials were disposed of in accordance with "Prudent Practices for Disposal of Chemicals from Laboratories"; National Academy Press; Washington, DC, 1983.

3. Discussion

Recent work has shown that a large variety of carbon, nitrogen, oxygen, sulfur, and halogen nucleophiles attack chiral N-protected serine β–lactones at the β–carbon to give optically pure N-protected α–amino acids.[2,4] However in certain cases (e.g., β–azidoalanine[5]) these products are unstable to most common deprotection conditions. The procedure given here describes the preparation of (S)-3-amino-2-oxetanone p-toluenesulfonic acid salt, a compound which reacts with a variety of nucleophiles to afford unprotected, optically pure α–amino acids directly (Scheme 1).[6] This salt has a long shelf life (many months) at room temperature provided that it is stored dry. It reacts rapidly with water ($t_{1/2}$ ~ 2.5 hr in unbuffered water; $t_{1/2}$ ~ 10 min in 50 mM potassium phosphate at pH 6.8). However, good nucleophiles such as thiols afford high yields of sulfur-containing amino acids in water if the pH is kept at 5.0-5.5.[6] Since both enantiomers of serine are relatively inexpensive, and L-serine is readily available in isotopically labeled form, this approach should prove useful for syntheses of sensitive D-amino acids as well as for preparation of the labeled L-isomers.

1. Department of Chemistry, University of Alberta, Edmonton, Alberta, Canada, T6G 2G2.
2. (a) Arnold, L. D.; Kalantar, T. H.; Vederas, J. C. *J. Am. Chem. Soc.* **1985**, *107*, 7105; (b) Arnold, L. D.; Drover, J. C. G.; Vederas, J. C. *J. Am. Chem. Soc.* **1987**, *109*, 4649.
3. Still, W. C.; Kahn, M.; Mitra, A. *J. Org. Chem.* **1978**, *43*, 2923.
4. See accompanying preparation of N$^\alpha$-(benzyloxycarbonyl)-β-(pyrazol-1-yl)-L-alanine. Pansare, S. V.; Huyer, G.; Arnold, L. D.; Vederas, J. C. *Org. Synth.* **1991**, *70*, 1.

5. Owais, W. M.; Rosichan, J. L.; Ronald, R. C.; Kleinhofs, A.; Nilan, R. A. *Mutat. Res.* **1983**, *118*, 229.
6. Arnold, L. D.; May, R. G.; Vederas, J. C. *J. Am. Chem. Soc.* **1988**, *110*, 2237.

Scheme 1

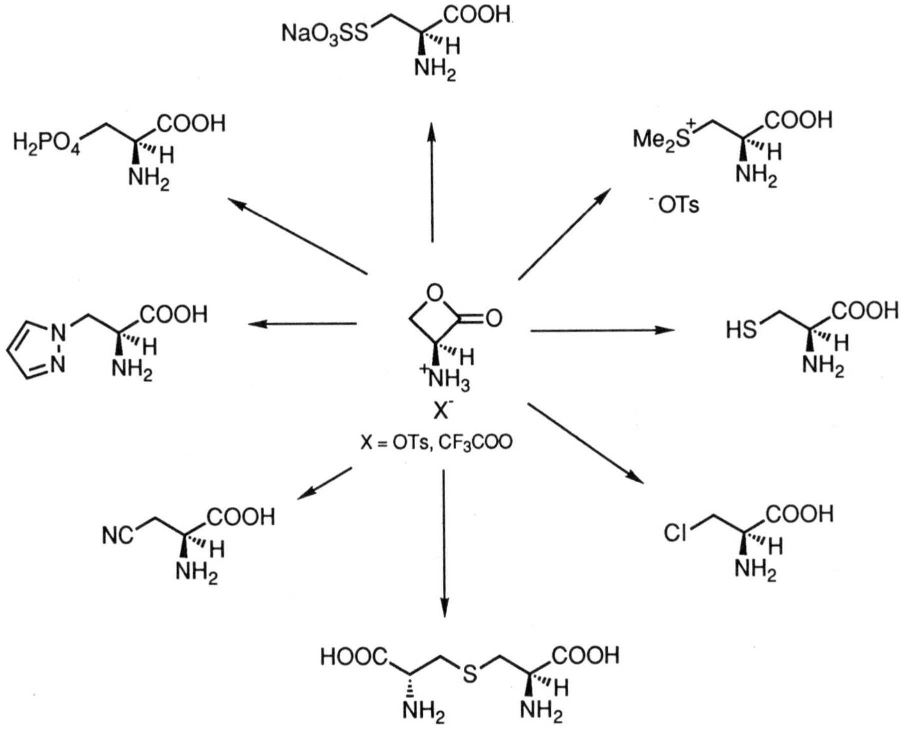

Appendix

Chemical Abstracts Nomenclature (Collective Index Number); (Registry Number)

N-tert-Butoxycarbonyl-L-serine β-lactone: Carbamic acid, (2-oxo-3-oxetanyl)-, 1,1-dimethylethyl ester, (S)- (11); (98541-64-1)

(S)-3-Amino-2-oxetanone p-toluenesulfoante: 2-Oxetanone, 3-amino-, (S)-, 4-methylbenzenesulfonate (12); (112839-95-9)

Triphenylphosphine: Phosphine, triphenyl- (8,9); (603-35-0)

Diethyl azodicarboxylate: Formic acid, azodi-, diethyl ester (8); Diazenedicarboxylic acid, diethyl ester (9); (1972-28-7)

N-tert-Butoxycarbonyl-L-serine: Serine, N-carboxy-, N-tert-butyl ester, L- (8); L-Serine, N[(1,1-dimethylethoxy)carbonyl]- (9); (3262-72-4)

p-Toluenesulfonic acid monohydrate (8); Benzenesulfonic acid, 4-methyl-, monohydrate (9); (6192-52-5)

Trifluoroacetic acid: Acetic acid, trifluoro- (8,9); (76-05-1)

1,1,-DIMETHYLETHYL (S)- OR (R)-4-FORMYL-2,2-DIMETHYL-3-OXAZOLIDINECARBOXYLATE: A USEFUL SERINAL DERIVATIVE

(3-Oxazolidinecarboxylic acid, 4-formyl-2,2-dimethyl-, 1,1-dimethyl ester, (S)- or (R)-)

A. HO-CH₂-*CH(NH₂)-CO₂H → [1. Boc₂O; 2. CH₃I] → HO-CH₂-*CH(NHBoc)-CO₂Me

B. HO-CH₂-*CH(NHBoc)-CO₂Me → [DMP, cat. TsOH] → oxazolidine-*CH(CO₂Me)-N-Boc

C. oxazolidine-*CH(CO₂Me)-N-Boc → [DIBAL, -78°C] → oxazolidine-*CH(CHO)-N-Boc

Submitted by Philip Garner and Jung Min Park.[1]
Checked by Mayumi Takasu and Hisashi Yamamoto.

1. Procedure

A. *N-[(1,1-Dimethylethoxy)carbonyl]-L-serine methyl ester.* A solution of di-tert-butyl dicarbonate [(Boc)₂O] (78.4 g, 0.36 mol, Note 1) in dioxane (280 mL, Note 2) is added to an ice-cold, magnetically stirred solution of L-serine (31.7 g, 0.30 mol, Note 3) in 1 N sodium hydroxide (620 mL) by means of an addition funnel. The two-phase mixture is stirred at 5°C for 30 min, then allowed to warm to room temperature over 3.5 hr at which time TLC analysis shows the reaction to be complete (Note 4). The mixture is concentrated to half its original volume by rotary evaporation at 35°C, cooled in an ice-water bath, acidified to pH 2-3 by the slow addition of 1 N potassium bisulfate (620 mL), and then extracted with ethyl acetate (3 x 1000 mL). The combined extracts are dried over magnesium sulfate, filtered and concentrated to give N-Boc-L-serine (63.0 g) as a colorless, sticky foam which is used without further purification.

To a cold solution of N-Boc-L-serine (32.4 g, 0.16 mol, Note 5) in dimethylformamide (150 mL) is added solid potassium carbonate (24.3 g, 0.176 mol). After stirring for 10 min in an ice-water bath, methyl iodide (20.0 mL, 46.3 g, 0.32 mol - *Caution! Methyl iodide is toxic and a suspected carcinogen that should be handled in a well-ventilated fume hood.*) is added to the white suspension and stirring continued at 0°C for 30 min whereupon the mixture solidifies. The reaction is warmed to room temperature and stirred for an additional hour or so at which point TLC analysis indicates complete formation of the methyl ester (Note 6). The reaction mixture is filtered by suction and the filtrate partitioned between ethyl acetate (300 mL) and water (300 mL). The organic phase is washed with brine (2 x 300 mL), dried with magnesium sulfate, filtered and concentrated to give 29.8 g (86% yield) of N-Boc-L-serine methyl ester as a pale amber oil which is used without further purification (Notes 7 and 8).[2]

B. *3-(1,1-Dimethylethyl) 4-methyl (S)-2,2-dimethyl-3,4-oxazolidinedicarboxylate.* To a 2-L, three-necked, round-bottomed flask, equipped with a magnetic stirring bar, Claisen distilling head, thermometer, and reflux condenser protected from moisture by a calcium sulfate-filled drying tube are added a solution of N-Boc-L-serine methyl ester (48.5 g, 0.22 mol) in benzene (770 mL), 2,2-dimethoxypropane (55 mL, 47 g, 0.45 mol), and p-toluenesulfonic acid monohydrate (0.593 g, 3.1 mmol, Note 9). The colorless solution is heated under reflux (oil bath temperature, 110°C) for 30 min, then slowly distilled until a volume of 660 mL is collected over 4 hr when the reaction is judged to be complete by TLC (Notes 10 and 11). The cooled, amber solution is partitioned between saturated sodium bicarbonate solution (200 mL) and ethyl ether

(2 x 500 mL). The organic layer is washed with brine (200 mL), then dried over magnesium sulfate, filtered and concentrated to give the crude product as an amber oil. This material is vacuum distilled through a 10-cm Vigreux column to give 40.3-50.9 g (70-89% yield) of oxazolidine methyl ester as a very pale yellow liquid, bp 101-102°C (2 mm) (Note 12).

  C. *1,1-Dimethylethyl (S)-4-formyl-2,2-dimethyl-3-oxazolidinecarboxylate.* A dry (Note 13), 1-L, three-necked, round-bottomed flask is equipped with a magnetic stirring bar, rubber septa, three-way balloon adapter, and a low temperature thermometer. After the flask is purged with nitrogen (Note 14), a solution of the oxazolidine ester (40.2 g, 0.15 mol) in dry toluene (300 mL, Note 15) is added via cannula (using positive nitrogen pressure) and cooled to -78°C with an acetone-dry ice bath. To this cooled solution is added a -78°C solution of 1.5 M diisobutylaluminum hydride in toluene (175 mL, Note 16) via cannula (using positive nitrogen pressure). The rate of addition is adjusted so as to keep the internal temperature below -65°C and takes approximately 1 hr to complete. The reaction mixture is stirred for an additional 2 hr at -78°C under an atmosphere of nitrogen when TLC anlaysis shows the reaction to be complete (Note 17). The reaction is quenched by slowly adding 60 mL of cold (-78°C) methanol (*Evolution of hydrogen occurs!*) - again so as to keep the internal temperature below -65°C. The resulting white emulsion is slowly poured into 1000 mL of ice-cold 1 N hydrochloric acid with swirling over 15 min, and the aqueous mixture is then extracted with ethyl acetate (3 x 1000 mL). The combined organic layers are washed with brine (1000 mL), dried over magnesium sulfate, filtered and concentrated to give 33.6 g of crude product as a colorless oil. This material is vacuum distilled through a 10-cm Vigreux column to give 26.9 g (76% yield) of oxazolidine aldehyde as a colorless liquid, bp 83-88°C (1.0-1.4 mm) (Note 18).

2. Notes

1. The submitters used di-tert-butyl dicarbonate purchased from Aldrich Chemical Company, Inc., also available from Wako Pure Chemical Industries, LTD.

2. Unless stated otherwise in the procedure, all solvents and reagents were used as purchased without further purification.

3. The submitters used L-serine (and D-serine) purchased from United States Biochemical Corporation, also available from Tokyo Kasei Kogyo Co., LTD.

4. TLC analysis on Merck silica gel 60F-254 plates eluting with (8:1:1) n-BuOH-H_2O-AcOH showed the clean formation of a product with R_f 0.65 (visualized with 0.3% ninhydrin in (97:3) n-BuOH-AcOH) at the expense of starting amino acid at the origin. If starting material remained, more $(Boc)_2O$ (13.1 g, 0.060 mol) was added.

5. The submitters have also found commercially available N-Boc-L-serine (United States Biochemical Corporation) to be an entirely satisfactory starting material. However, for the unnatural D-series, they find that it is more economical to prepare N-Boc-D-serine as described.

6. TLC analysis on Merck silica gel 60F-254 plates eluting with (1:1) ethyl acetate-hexanes showed the clean formation of ester, R_f 0.38 (visualized with 0.5% phosphomolybdic acid in 95% ethanol), at the expense of starting material at the origin.

7. Optical measurements on this material were not very useful since they were in general low and quite variable. Furthermore, no literature values could be found for comparison. IR (neat) cm^{-1}: 3400, 1720 (br); ^{1}H NMR (200 MHz, C_6D_6, 17°C) δ: 1.41 (s, 9 H), 2.50 (br, s, H, exchanged with D_2O), 3.26 (s, 3 H), 3.66 (dd, H, J = 11 and 4), 3.76 (dd, H, J = 11 and 4), 4.40 (m, H), 5.60 (m, H, exchanged with D_2O).

8. *Alternatively, the methyl ester could be prepared as follows:* N-Boc-serine (63 g, 0.31 mol) was dissolved in ethyl ether (600 mL) in a 2-L Erlenmeyer flask equipped with a magnetic stirring bar, cooled in an ice-water bath and treated with ten 50-mL aliquots of cold ethereal (approximately 0.6 M) diazomethane prepared from N-nitroso-N-methylurea according to Arndt's procedure.[3] After 30 min at 0°C, TLC analysis showed the reaction to be complete (Note 6). Excess diazomethane was destroyed with acetic acid (the yellow color disappears) and the resulting solution was extracted with half-saturated sodium bicarbonate solution (300 mL), then washed with brine (200 mL), dried with magnesium sulfate, filtered and concentrated to give 60.1 g (91% over 2 steps) of N-Boc-serine methyl ester as a colorless, sticky foam which was used without further purification. *Caution! N-Nitroso-N-methylurea is suspected of being a carcinogen and diazomethane is highly toxic. The utmost care must be used when handling these substances; diazomethane solutions should be restricted to a well-ventilated fume hood at all times.*

9. Moriwake et al. have reported that boron trifluoride etherate can also be used as the acid catalyst for this reaction.[6]

10. TLC analysis on Merck silica gel 60F-254 plates eluting with (1:1) ethyl acetate-hexanes showed the clean formation of product, R_f 0.78 (visualized with 0.5% phosphomolybdic acid in 95% ethanol), at the expense of starting material at R_f 0.23.

11. If starting material remained at this time, more 2,2-dimethoxypropane (14 mL, 12 g, 0.11 mol) and benzene (310 mL) were added and the procedure was repeated, collecting 250 mL of distillate, at which time the TLC analysis generally showed the reaction to be complete.

12. The optical rotation of the L-oxazolidine methyl ester was -46.7° (CHCl$_3$, *c* 1.30). An essentially identical procedure emanating from N-Boc-D-serine methyl ester gave the corresponding D-oxazolidine methyl ester in 80% yield with a rotation of +53°. In either case further purification could be achieved with flash chromatography

to give product with a maximum rotation of 57° although we have found distilled material to be entirely satisfactory for our purposes: IR (neat) cm^{-1}: 1760, 1704; ^{1}H NMR (200 MHz, C$_6$D$_6$, 75°C) δ: 1.41 (s, 9 H), 1.53 (br s, 3 H), 1.81 (br s, 3 H), 3.35 (s, 3 H), 3.75 (dd, H, J = 8.5 and 8.1), 3.81 (dd, H, J = 8.5 and 3.5), 4.26 (m, H). (The oxazolidine derivatives exist as slowly interconverting rotamers on the NMR time scale and samples require heating to obtain averaged spectra.)

13. All the glassware (except the low-temperature thermometer) was oven-dried (>100°C) and quickly assembled before use.

14. The checkers used argon.

15. Toluene was distilled from sodium-benzophenone ketyl.

16. The diisobutylaluminum hydride solution (1.5 M in toluene, Aldrich Chemical Company, Inc.) was transferred to a dry, 250-mL, graduated cylinder equipped with a rubber septum and drying tube via cannula (using positive nitrogen pressure). The graduated cylinder was then placed in a Dewar flask and cooled to -78°C with an acetone-dry ice bath.

17. TLC analysis on Merck silica gel 60F-254 plates eluting with (4:1) hexanes-ethyl acetate showed the formation of product, R$_f$ 0.33 (visualized with 0.5% phosphomolybdic acid in 95% ethanol), with only a trace of starting material remaining at R$_f$ 0.41. Some over-reduced product arising within the TLC capillary may be evident at this stage.

18. The optical rotation of the L-oxazolidine aldehyde was -91.7° (CHCl$_3$, c 1.34). An identical procedure emanating from the D-oxazolidine methyl ester gave D-oxazolidine aldehyde in 85% yield having a rotation of +95°. These distilled products contained up to 5% of the starting material as judged by their NMR spectra, but were suitable for use without further purification. Homogeneous samples could be obtained in either case by flash chromatography on silica gel eluting with (4:1) hexanes-ethyl acetate and showed a maximum optical rotation of 105°. This material can be stored

indefinitely provided it is kept cold (≤ 5°C) and moisture-free. IR (neat) cm^{-1}: 1735, 1705; ^{1}H NMR (200 MHz, C$_6$D$_6$, 60°C) δ: 1.34 (s, 9 H), 1.40 (br s, 3 H), 1.59 (br s, 3 H), 3.52 (dd, H, J = 8.7 and 8.3), 3.65 (dd, H, J = 8.7 and 2.9), 3.90 (m, H), 9.34 (br s, H).

Waste Disposal Information

All toxic materials were disposed of in accordance with "Prudent Practices for Disposal of Chemicals from Laboratories"; National Academy Press; Washington, DC, 1983.

3. Discussion

Since its preparation and use was first reported by us,[4] the title compound has been gaining favor as a chiral, nonracemic synthon for the asymmetric synthesis of a variety of amino alcohol- and amino acid-containing targets. Among the virtues of this oxazolidine aldehyde over previously reported N-acylated serinal derivatives are its ease of preparation on a large scale and its configurational stability. The procedure described here provides material that has been determined to be 95% enantiomerically pure by Mosher ester analysis.[7] Homologation (C-C bond formation) of this serinal derivative can be achieved without competing racemization using both olefination[6,15,22-26] and nucleophilic addition[4,5,8-14,16-21] protocols. The latter process can be made to occur with good to excellent diastereoselectivity (i.e., 1,2-asymmetric induction) by simply choosing reagents/conditions so as either to preclude or favor chelation-control. Protocols for diastereoselective additions to the oxazolidine appended olefins are also known. Once the rest of the target molecule's structure is in place, the oxazolidine can be unravelled to give either an aminoethanol group or, after oxidation of the primary alcohol, an α–amino acid moiety. The products so produced are typically >98% enantiomerically pure. Thus the title compound can serve not only as a serinal derivative but as a penaldic acid equivalent as well. Representative examples of the use of 1,1-dimethylethyl (S)- and (R)-4-formyl-2,2-dimethyl-3-oxazolidinecarboxylate as a chiral synthon for natural product synthesis are collected in Table I.

1. Department of Chemistry, Case Western Reserve University, Cleveland, OH 44106-7078.
2. We thank Dr. N. S. Chandrakumar of G. D. Searle and Company for alerting us to the fact that N-Boc-serine methyl ester could be prepared without significant racemization via esterification with methyl iodide, thereby avoiding the generation and use of diazomethane on a large scale.
3. Arndt, F. *Org. Synth., Coll. Vol. II* **1943**, 165.
4. Garner, P. *Tetrahedron Lett.* **1984**, *25*, 5855.
5. Garner, P.; Ramakanth, S. *J. Org. Chem.* **1986**, *51*, 2609; Garner, P.; Ramakanth, S.; Yoo, J. U., unpublished results.
6. Moriwake, T.; Hamano, S.; Saito, S.; Torri, S. *Chem. Lett.* **1987**, 2085.
7. Garner, P.; Park, J. M. *J. Org. Chem.* **1987**, *52*, 2361.
8. Dondoni, A.; Fantin, G.; Fogagnolo, M.; Medici, A. *J. Chem. Soc., Chem. Commun.* **1988**, 10; Dondoni, A.; Fantin, G.; Fogagnolo, M.; Pedrini, P. *J. Org. Chem.* **1990**, *55*, 1439.
9. Herold, P. *Helv. Chim. Acta* **1988**, *71*, 354.
10. Nimkar, S.; Menaldino, D.; Merrill, A. H.; Liotta, D. *Tetrahedron Lett.* **1988**, *29*, 3037.
11. Garner, P.; Park, J. M. *J. Org. Chem.* **1988**, *53*, 2979.
12. Garner, P.; Park, J. M.; Malecki, E. *J. Org. Chem.* **1988**, *53*, 4395.
13. Radunz, H. E.; Devant, R. M.; Eiermann, V. *Liebigs Ann. Chem.* **1988**, 1103.

14. Casiraghi, G.; Cornia, M.; Rassu, G. *J. Org. Chem.* **1988**, *53*, 4919.
15. Yanagida, M.; Hashimoto, K.; Ishida, M.; Shinozaki, H.; Shirahama, H. *Tetrahedron Lett.* **1989**, *30*, 3799.
16. Garner, P.; Park, J. M. *Tetrahedron Lett.* **1989**, *29*, 5065; Garner, P.; Park, J. M. *J. Org. Chem.* **1990**, *55*, 3772.
17. Casiraghi, G.; Colombo, L.; Rassu, G.; Spanu, P. *Tetrahedron Lett.* **1989**, *30*, 5325; Casiraghi, G.; Colombo, L.; Rassu, G.; Spanu, P.; Fava, G. G.; Belicchi, M. F. *Tetrahedron* **1990**, *46*, 5807.
18. Dondoni, A.; Fantin, G.; Fogagnolo, M. *Tetrahedron Lett.* **1990**, *31*, 6063.
19. Kahne, D.; Yang, D.; Lee, M. D. *Tetrahedron Lett.* **1990**, *31*, 21.
20. Nakagawa, M.; Tsuruoka, A.; Yoshida, J.; Hino, T. *J. Chem. Soc., Chem. Commun.* **1990**, 603.
21. Dondoni, A.; Fantin, G.; Fogagnolo, M.; Merino, P. *J. Chem. Soc., Chem. Commun.* **1990**, 854.
22. Priepke, H.; Brückner, R.; Harms, K. *Chem. Ber.* **1990**, *123*, 555.
23. Jako, I.; Uiber, P.; Mann, A.; Taddei, M.; Wermuth, C-G. *Tetrahedron Lett.* **1990**, *31*. 1011.
24. Chung, J. Y. L.; Wasicak, J. T. *Tetrahedron Lett.* **1990**, *31*, 3957.
25. Shimamoto, K.; Ohfune, Y. *Tetrahedron Lett.* **1990**, *31*, 4049.
26. Sakai, N.; Ohfune, Y. *Tetrahedron Lett.* **1990**, *31*, 4151.

TABLE

SELECTED EXAMPLES OF THE USE OF 1,1-DIMETHYL (S)- AND (R)-4-FORMYL-2,2-DIMETHYL-3-OXAZOLIDINECARBOXYLATE FOR NATURAL PRODUCT SYNTHESIS: THE C-3 SUBUNITS EMANATING FROM SERINE ARE NUMBERED ACCORDINGLY

threo-β-hydroxy-L-glutamic acid (ref. 4)

D-*erythro*-sphingosine (refs. 8-10,12,13)

5-carbamoylpolyoxamic acid (ref. 11)

Amipurimycin studies (ref. 5)

Thymine polyoxin C (ref. 16)   Calicheamycin fragment (ref. 19)   (-)-Galantic acid (ref. 26)

Appendix

Chemical Abstracts Nomenclature (Collective Index Number); (Registry Number)

1,1-Dimethylethyl (S)- or (R)-4-formyl-2,2-dimethyloxazolidinecarboxylate: 3-Oxazolidinecarboxylic acid, 4-formyl-2,2-dimethyl-, 1,1-dimethylethyl ester, (S)- or (R)- (11); (S)- (102308-32-7); (R)- (95715-87-0)

N-[(1,1-Dimethylethoxy)carbonyl]-L-serine methyl ester: L-Serine, N-[(1,1-dimethylethoxy)carbonyl]-, methyl ester (9); (2766-43-0)

Di-tert-butyl dicarbonate: Formic acid, oxydi-, di-tert-butyl ester (8); Dicarbonic acid, bis(1,1-dimethylethyl) ester (9); (24424-99-5)

Serine: L-Serine (8,9); (56-45-1)

N-tert-Butoxycarbonylserine: Serine, N-carboxy-, N-tert-butyl ester, L- (8); L-Serine, N-[(1,1-dimethylethoxy)carbonyl]- (9); (3262-72-4)

Methyl iodide: Methane, iodo- (8,9); (74-88-4)

3-(1,1-Dimethylethyl) 4-methyl (S)-2,2-dimethyl-3,4-oxazolidinedicarboxylate: 3,4-Oxazolidinedicarboxylic acid, 2,2-dimethyl-, 3-(1,1-dimethylethyl) 4-methyl ester, (S)- (12); (108149-60-6)

p-Toluenesulfonic acid monohydrate (8) Benzenesulfonic acid, 4-methyl-, monohydrate (9); (6192-52-5)

Diisobutylaluminum hydride: Aluminum, hydrodiisobutyl- (8); Aluminum, hydrobis(2-methylpropyl)- (9); (1191-15-7)

Nitrosomethylurea: Urea, N-methyl-N-nitroso-; (684-93-5)

N-(BENZYLOXYCARBONYL)-L-VINYLGLYCINE METHYL ESTER

(3-Butenoic acid, 2-[[(phenylmethoxy)carbonyl]amino]-, methyl ester, (S)-)

A. [Structure 1: methionine methyl ester hydrochloride] $\xrightarrow{\text{CbzCl}, K_2CO_3}$ [Structure 2: N-Cbz methionine methyl ester]

B. [Structure 2] $\xrightarrow{\text{NaIO}_4}$ [Structure 3: sulfoxide]

C. [Structure 3] $\xrightarrow{\Delta}$ [Structure 4: vinylglycine derivative]

Submitted by Michael Carrasco, Robert J. Jones, Scott Kamel, H. Rapoport,[1] and Thien Truong.

Checked by Antje Grützmann and Ekkehard Winterfeldt.

1. Procedure

A. N-(Benzyloxycarbonyl)-L-methionine methyl ester (**2**). A 3-L, three-necked, Morton flask equipped with an efficient mechanical stirrer, thermometer, and a dropping funnel is charged with L-methionine methyl ester hydrochloride (117.6 g, 0.56 mol) (Note 1), potassium bicarbonate (282.3 g, 2.82 mol, 500 mol %), water (750 mL), and ether (750 mL), and the solution is cooled to 0°C. Benzyl chloroformate (105 g, 88.6 mL, 0.62 mol, 110 mol %, Aldrich Chemical Company, Inc.) is added dropwise over 1 hr, the cooling bath is removed, and the solution is stirred for 5 hr. Glycine (8.5 g, 0.11 mol, 20 mol %, Aldrich Chemical Company, Inc.) is added (to scavenge excess chloroformate) and the solution is stirred for an additional 18 hr. The organic layer is separated, and the aqueous layer is extracted with ether (2 x 200 mL). The combined organic layers are washed with 0.01 M hydrochloric acid (2 x 500 mL), water (2 x 500 mL), and saturated brine (500 mL), and then dried (Na_2SO_4), filtered, and evaporated on a rotary evaporator. The resulting oil is further dried in a Kugelrohr oven (50°C, 0.1 mm, 12 hr) to leave product **2** as a clear oil that solidifies upon cooling: 165-166 g (98-99%), mp 42-43°C.

    B. Methyl L-2-(benzyloxycarbonylamino)-4-(methylsulfinyl)butanoate (**3**). A 5-L, three-necked, Morton flask equipped with an efficient mechanical stirrer, thermometer, and dropping funnel is charged with **2** (166.0 g, 0.56 mol) and methanol (1.5 L), and the solution is cooled to 0°C. A solution of sodium periodate ($NaIO_4$) (131.4 g, 0.61 mol, 110 mol %) in water (2 L) is added dropwise over a period of 1.5 hr. The cooling bath is removed and the mixture is stirred for 18 hr. The product is vacuum-filtered through Celite and divided into two portions. Each portion is extracted with chloroform (6 x 200 mL), washed with water (300 mL) and brine (300 mL), dried (Na_2SO_4), filtered, and evaporated by rotary evaporation (bath temperature <30°C). The resulting oils are combined and further dried in a Kugelrohr oven (30°C, 0.1 mm, 12 hr), yielding the product as a waxy solid: 173.2 g, 99%.

    C. N-(Benzyloxycarbonyl)-L-vinylglycine methyl ester (**4**). Sulfoxide **3** (35.0 g, 0.11 mol) and Pyrex helices (35 g) are placed in a 1-L, round-bottomed flask, thoroughly mixed by shaking, and distilled from a rocking Kugelrohr apparatus (195-200°C, 0.1-0.3 mm, 1 hr) into a chilled receiving flask cooled in powdered dry ice to afford a yellow oil (Notes 2 and 3). Low pressure chromatography (LPC) of the crude oil gives the N-protected vinylglycine methyl ester **4** (17.4 g, 62%) of 95% purity (Notes

4 and 5). Medium pressure liquid chromatography (MPLC) of the crude oil provides pure **4** in 60% yield from **3** (Note 6). L-Vinylglycine hydrochloride can be obtained from **4** in almost quantitative yield by refluxing in 6 N hydrochloric acid for 1 hr.[2]

2. Notes

1. L-Methionine methyl ester hydrochloride is commercially available (Aldrich Chemical Company, Inc.); however, it is prepared easily as follows: A 3-L, three-necked, Morton flask is equipped with an efficient mechanical stirrer. The flask is charged with L-methionine (100.0 g, 0.67 mol) and methanol (0.7 L), the solution is cooled to 0°C, and hydrogen chloride gas is bubbled through the mixture for 15 min (in about 2 min the solution becomes homogeneous). The cooling bath is removed, the solution is stirred for 18 hr, and the solvent is evaporated. Further drying under reduced pressure gives L-methionine methyl ester hydrochloride as a white solid (132.5 g, 99%), that is suitable for most purposes. It can be recrystallized by dissolving in hot methanol (500 mL) and precipitating with ether (1 L) to give the pure hydrochloride: 117.6 g, 88%, mp 152-153°C.

2. *Caution: Stench. The entire reaction apparatus -- Kugelrohr oven, vacuum pump, and subsequent chromatography -- should be kept in an efficient fume hood.*

3. TLC of the distillate shows **4** (2/1, hexanes/ethyl acetate; visualization by staining with 5% ethanolic molybdophosphoric acid and charring) as the major product (R_f 0.37) along with minor amounts of the (E)- and (Z)-α,β-unsaturated isomer (R_f 0.40 and 0.30).

4. LPC conditions are as follows: 9-cm diameter column; 800 mL of 230-400 mesh EM Science silica gel; 4/1, hexanes/ethyl acetate (1.8 L) to 2/1, hexanes/ethyl acetate.

5. The ^{1}H NMR spectrum of **4** is as follows: (CDCl$_3$) δ: 3.77 (s, 3 H, CO$_2$CH$_3$), 4.94 (m, 1 H, α-H), 5.13 (s, 2 H, CH$_2$PH), 5.28 (dd, 1 H, J = 1.2, 10.3, H$_{cis}$), 5.36 (dd, 1 H, J = 1.4, 17.1, H$_{trans}$), 5.47 (bd, 1 H, NH), 5.91 (m, 1 H, H$_{vinyl}$), 7.35 (s, 5 H, ArH).

6. MPLC conditions are as follows: 40 cm x 6-cm column; 230-400 mesh EM Science silica gel; flow rate 18 mL/min; model 153 Altex UV Detector; retention time ca. 100 min; 4/1, hexanes/ethyl acetate. The distillate was chromatographed in 2.5-g batches. The checkers experienced significant losses in this step, which may be highly sensitive to the type of silica and the apparatus used. In any case the material obtained after the first chromatography will be acceptable for most purposes.

Waste Disposal Information

All toxic materials were disposed of in accordance with "Prudent Practices for Disposal of Chemicals from Laboratories"; National Academy Press; Washington, DC, 1983.

3. Discussion

Vinylglycine is a natural amino acid found in mushrooms.[3] It is an inhibitor of pyridoxal-linked aspartate aminotransferase,[4] and has also been postulated as an intermediate in the enzymatic conversion of homoserine to threonine[5] and α-ketobutyrate.[6] Protected vinylglycine is also a versatile asymmetric starting material for synthesis.[7] Variants have been prepared in racemic,[8-13] optically active,[14] optically pure,[2,15-17] and isotopically labeled form.[4b,18-20] This procedure is derived from our earlier publication[2] and contains improvements in procedure and scale-up.

1. Department of Chemistry, University of California, Berkeley, CA 94720.
2. Afzali-Ardakani, A.; Rapoport, H. *J. Org. Chem.* **1980**, *45*, 4817.
3. Dardenne, G.; Casimir, J.; Marlier, M.; Larsen, P. O. *Phytochemistry* **1974**, *13*, 1897.
4. (a) Rando, R. R.; Relyea, N.; Cheng, L. *J. Biol. Chem.* **1976**, *251*, 3306; (b) Rando, R. R. *Biochemistry* **1974**, *13*, 3859.
5. Flavin, M.; Slaughter, C. *J. Biol. Chem.* **1960**, *235*, 1112.
6. Posner, B. I.; Flavin, M. *J. Biol. Chem.* **1972**, *247*, 6402.
7. Shaw, K. J.; Luly, J. R.; Rapoport, H. *J. Org. Chem.* **1985**, *50*, 4515.
8. Friis, P.; Helboe, P.; Larsen, P. O. *Acta Chem. Scand., Ser. B* **1974**, *28*, 317.
9. Baldwin, J. E.; Haber, S. B.; Hoskins, C.; Kruse, L. I. *J. Org. Chem.* **1977**, *42*, 1239.
10. Hudrlik, P. F.; Kulkarni, A. K. *J. Am. Chem. Soc.* **1981**, *103*, 6251.
11. Vyas, D. M.; Chiang, Y.; Doyle, T. W. *J. Org. Chem.* **1984**, *49*, 2037.
12. Greenlee, W. *J. Org. Chem.* **1984**, *49*, 2632.
13. Fitzner, J. N.; Pratt, D. V.; Hopkins, P. B. *Tetrahedron Lett.* **1985**, *26*, 1959.
14. Schöllkopf, U.; Nozulak, J.; Groth, U. *Tetrahedron* **1984**, *40*, 1409.
15. Hanessian, S.; Sahoo, S. P. *Tetrahedron Lett.* **1984**, *25*, 1425.
16. Barton, D. H. R.; Crich, D.; Hervé, Y.; Potier, P.; Thierry, J. *Tetrahedron* **1985**, *41*, 4347.
17. Pellicciari, R.; Natalini, B.; Marinozzi, M. *Synth. Commun.* **1988**, *18*, 1715.
18. Chang, M. N. T.; Walsh, C. T. *J. Am. Chem. Soc.* **1981**, *103*, 4921.
19. Sawada, S.; Nakayama, T.; Esaki, N.; Tanaka, H.; Soda, K.; Hill, R. K. *J. Org. Chem.* **1986**, *51*, 3384.
20. Rosegay, A.; Taub, D. *Synth. Commun.* **1989**, *19*, 1137.

Appendix

Chemical Abstracts Nomenclature (Collective Index Number); (Registry Number)

N-(Benzyloxycarbonyl)-L-vinylglycine methyl ester: 3-Butenoic acid, 2-[[(phenylmethoxy)carbonyl]amino]-, methyl ester, (S)- (10); (75266-40-9)

N-(Benzyloxycarbonyl)-L-methionine methyl ester: L-Methionine, N-[(phenylmethoxy)carbonyl]-, methyl ester (9): (56762-93-7)

L-Methionine methyl ester hydrochloride: Methionine, methyl ester, hydrochloride, L- (8); L-Methionine, methyl ester, hydrochloride (9); (2491-18-1)

Benzyl chloroformate: Formic acid, chloro-, benzyl ester (8); Carbonochloride acid, phenylmethyl ester (9); (501-53-1)

Glycine (8,9); (56-40-6)

Methyl L-2-(benzyloxycarbonylamino)-4-(methylsulfinyl)butanoate: Butanoic acid, 4-(methylsulfinyl)-2-[[(phenylmethoxy)carbonyl]amino]-, methyl ester, (S)- (10); (75266-39-3)

L-Vinylglycine hydrochloride: 3-Butenoic acid, 2-amino-, hydrochloride, (S)- (10); (75266-38-5)

L-Methionine (8,9); (63-68-3)

(R)-(-)-10-METHYL-1(9)-OCTAL-2-ONE

(2(3H)-Naphthalenone, 4,4a,5,6,7,8-hexahydro-4a-methyl-, (R)-)

Submitted by G. Revial and M. Pfau.[1]
Checked by Graham N. Maw and Robert K. Boeckman, Jr.

1. Procedure

A. *Imines* (1). In a 1000-mL, round-bottomed flask, fitted with a toluene-filled Dean-Stark water separator, 100.0 g (0.825 mol) of (S)-(-)-α-methylbenzylamine (Note 1), 92.5 g (0.825 mol) of 2-methylcyclohexanone (Note 2) and 100 mL of toluene are heated at reflux temperature under a nitrogen atmosphere. After 24 hr, ca. 15 mL of water (ca. 100% theor.) has been removed azeotropically.

B. *(R)-(+)-2-Methyl-2-(3-oxobutyl)cyclohexanone* (**2**). The solution of imines **1** is cooled in an ice bath and 72.5 mL (61.0 g, 0.870 mol, ca. 1.05 equiv) of freshly distilled methyl vinyl ketone (Note 3) is added with a syringe, with magnetic stirring, under a nitrogen atmosphere. The flask is then heated at ca. 40°C for 24 hr.

The slightly yellow solution is cooled in an ice bath and 60 mL of glacial acetic acid (ca. 1 mol) and 50 mL of water are added. Hydrolysis is achieved by stirring the heterogeneous mixture at room temperature for 2 hr. The now clear solution is poured into a 2000-mL separatory funnel containing 100 mL of brine and 160 mL of water and is extracted five times with a 50:50 mixture of ether-petroleum ether (35-60°C) (1000-mL total amount). The organic phase is washed efficiently with 20 mL of 10% hydrochloric acid, 20 mL of water, and two 10-mL portions of brine.

The aqueous layer is kept for recovery of the amine (see D below).

The pale yellow organic layer is dried over a small amount of anhydrous magnesium sulfate and filtered. The solvents are removed with a rotary evaporator at ca. 40°C and the crude diketone **2** (ca. 145 g) is used directly for the next step.

C. *(R)-(-)-10-Methyl-1(9)-octal-2-one* (**3**). Dry methanol (600 mL) is added to the crude diketone **2** contained in a 2000-mL, round-bottomed flask and a rubber septum is fitted. The solution is stirred at room temperature for 15 min under a slight stream of nitrogen to remove traces of oxygen. A 25 wt. % solution of sodium methoxide in methanol (Note 4) is then introduced dropwise with a syringe, under stirring, until a slightly red color develops (Note 5). At this point, 15 mL (ca. 0.07 mol) of the sodium methoxide solution is added and the mixture is heated at 60°C for 10 hr under a nitrogen atmosphere. The solution is cooled and the now deep red solution is neutralized with glacial acetic acid (ca. 4.5 mL) until the color turns yellow. Methanol is removed with a rotary evaporator until a thick paste (sodium acetate) results; this is dissolved with 200 mL of water. Extraction is effected in a 2000-mL separatory funnel using four portions of a 50:50 mixture (1000 mL total amount) of ether-petroleum ether (35-60°C). The pale orange organic phase is washed twice with 40 mL of water, then twice with 20 mL of brine. The organic layer is dried over a small amount of anhydrous magnesium sulfate and filtered. The solvents are removed with a rotary evaporator at ca. 40°C and the red oily residue of crude octalone **3** is distilled under reduced pressure to afford ca. 110 g of a colorless oil, bp 70°C (2 mm) (Note 6). The oil is purified by recrystallization at a low temperature with an apparatus designed for this purpose.[2] The oil is dissolved with stirring in a 1000-mL, round-bottomed flask with 370 mL of pentane under a nitrogen atmosphere. The flask is then cautiously cooled in a Dewar flask containing liquid nitrogen until the solution is solid. The Dewar flask is removed and the temperature is allowed to rise, while magnetic stirring is resumed. When a crystalline mass in suspension appears suddenly (Note 7), an acetone bath adjusted to about -10°C is installed. When only a few crystals remain in suspension, the bath temperature is slowly lowered to -35°C, inducing crystallization. After the mixture has been kept at -35°C for 15 min, magnetic stirring is stopped and efficient filtration is performed through the filter stick. Still at -35°C, the compound is washed with 40 mL of pentane. The solution is stirred vigorously for 5 min and filtered as before. The washing operation is repeated two more times. The flask is warmed to room temperature and the oil is distilled as before, bp 70°C (2 mm), to afford 60-65 g (44-48% overall yield from 2-methylcyclohexanone) of methyloctalone **3** [>99% chemical purity, capillary GLC (Note 6)], $[\alpha]_D^{20}$ -210° (ethanol, *c* 1.00), optical purity 96% (Note 8).

D. *Recovery of (S)-(-)-α–methylbenzylamine.* The aqueous layer (see above, B, 3rd paragraph) contained in a 1000-mL flask is cooled in an ice bath under a nitrogen atmosphere. An aqueous 10% sodium hydroxide solution is added with stirring until a pH of 12-14 (Note 9) is reached. The mixture is transferred to a

2000-mL separatory funnel and extracted three times with ether (1000-mL total amount). The organic phase is washed twice with 100 mL of water and twice with 30 mL of brine, then dried over potassium carbonate and filtered. The solution is concentrated at room temperature with a rotary evaporator and distilled under reduced pressure to afford 85-90 g (85-90% yield) of the amine, bp 70°C (15 mm), the specific rotation of which is the same as that of the original starting material.

2. Notes

1. (S)-(-)-α–Methylbenzylamine (95.8% optical purity) and its (R)-(+)-enantiomer (93.4% optical purity) were purchased from Aldrich Chemical Company, Inc. and used without further purification. *Caution: These amines are toxic.* Air must be excluded to prevent formation of carbonates.

2. 2-Methylcyclohexanone was purchased from Aldrich Chemical Company, Inc., and used without further purification.

3. Methyl vinyl ketone, purchased from Aldrich Chemical Company, Inc., is dried over anhydrous potassium carbonate for 0.5 hr. The now slightly-colored oil is filtered and distilled under a nitrogen atmosphere at reduced pressure, bp 55°C (260 mm). *The colorless center of the distillate is collected over a few milligrams (ca. 0.05% concentration) of hydroquinone.*

4. A 25 wt % solution of sodium methoxide in methanol (4.6 M) was purchased from Aldrich Chemical Company, Inc.

5. Any acetic acid present is thus neutralized (a few drops are usually sufficient). Confirm that a basic pH has been reached by testing a drop of the solution, removed by syringe.

6. Although a TLC test (silica gel Merck 60 F$_{254}$, ethyl acetate/hexane 20:80 elution, UV and chromic-sulfuric acid visualization) of this oil reveals only one spot (R$_f$ 0.6), capillary GLC (Hewlett-Packard HP-5, 25 m x 0.2 mm x 0.5 µm, 200°C) shows in addition to methyloctalone **3** (retention time, 10.9 min), an 8% impurity (retention time, 11.7 min) which is probably the regioisomer **4**.

4

A chemically pure sample of methyloctalone **3**, $[\alpha]_D^{20}$ -190° (ethanol, c 1.00) is obtained by preparative GLC (3 m x 1/4", 5% carbowax 20 M on Aeropak 30, 180°C, retention time 12.4 min) of a distillate sample. The impurity has a 13.2-min retention time under these conditions. A good approximation of the reaction ee is calculated as follows, taking into account the optical purity of the starting amine (95.8%) and the $[\alpha]_D^{20}$ -219° (ethanol, c 1.00) value (Note 8) for optically pure methyloctalone **3**: ee = 100 x 190 x 100/95.8 x 219 ≅ 91%.

7. If no crystalline mass is formed but instead an emulsion of the two liquid phases appears, the entire solidification must be repeated (liquid nitrogen cooling).

8. Optically pure methyloctalone **3** is obtained by recrystallizing a sample of the purified compound several times until a constant $[\alpha]_D^{20}$ -219° (ethanol, c 1.00 or methanol, c 1.10) is observed. The $[\alpha]_D^{20}$ -219° (methanol, c 1.1) value from the literature[3] is confirmed.

9. If a pH lower than 12 is used, recovery of the amine is poor.

Waste Disposal Information

All toxic materials were disposed of in accordance with "Prudent Practices for Disposal of Chemicals from Laboratories"; National Academy Press; Washington, DC, 1983.

3. Discussion

Optically active methyloctalone **3** (or its enantiomer) has been obtained by resolution[3,4] and by asymmetric synthesis.[5] Both enantiomers are also available commercially. The synthesis described here is by far the most simple and gives the best yield. It is an application of the general method reported for the enantioselective elaboration of quaternary carbon centers through Michael-type alkylation of chiral imines.[6]

This general method has the following advantages: very simple procedures, very mild conditions involving moderate reaction temperatures and no acids, bases, catalysts, etc., allowing the use of labile reactants, excellent regio- and enantioselectivities as well as high chemical yields, and creation of useful chiral building blocks with quaternary carbon centers involving functionalized chains, allowing one to devise syntheses of natural compounds of various types (terpenes, steroids, alkaloids, etc.). As both optically active amines are commercially available, targets with either desired absolute configuration can be synthesized. Easy recovery of the auxiliary chiral moiety is possible.

Racemic methyloctalone **3** has been used as a starting material for many transformations as well as for syntheses of racemic natural compounds. In the following examples, the absolute configuration of the methyloctalone that would be required to obtain the natural enantiomer is indicated in parentheses: β–gorgonene[7] (S), widdrol[8,9] (R), thujopsene[8,10] (R), (R), 7-hydroxycostal[11,12] (R), β-selinene[13,14]

(R), costol, costal and costic acid[13] (R), norketoagarofuran[15,16] (R), β-eudesmol[13,14,15,17] (R), taxane model[18] (S).

The following, chiral, key intermediates have been prepared by a procedure similar to the one described above, generally with excellent chemical and optical yields:

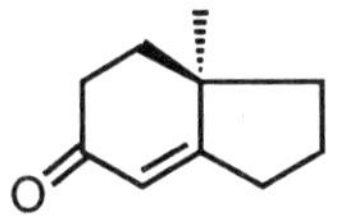

(R)[6a,b]

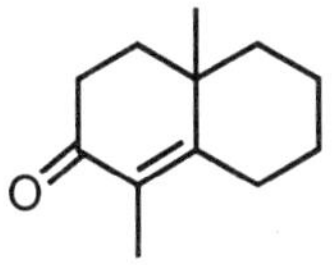

5 : (R) and (S)[19]

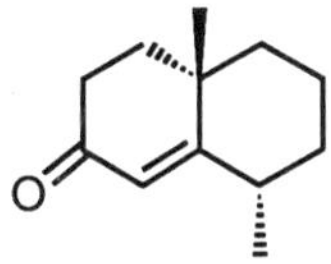

6 : (S,S)[19]

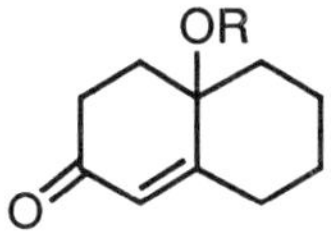

R = Me[20]
R = CH_2Ph (R)[20]

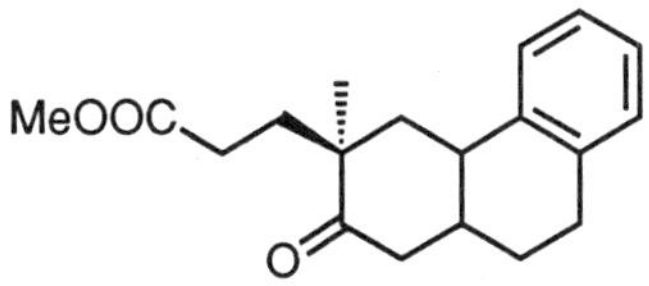

(R)[21]

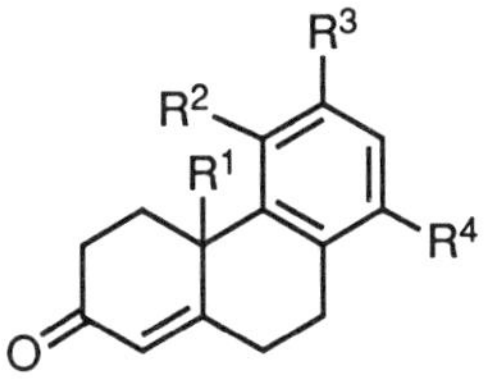

R^1 = Me; R^2 = R^3 = R^4 = H (R)[21]
7: R^1 = Me; R^2 = OMe; R^3 = H; R^4 = $(CH_2)_2COOH$ (S)[21]
8: R^1 = Me; R^2 = R^3 = H; R^4 = OMe (S)[22]
9: R^1 = CH_2COO-t-Bu; R^2 = R^4 = H; R^3 = OMe (S)[23]

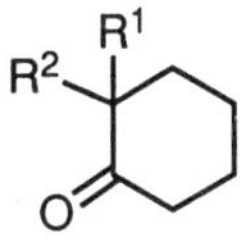

10: R^1 = Me; R^2 = $(CH_2)_2COOMe$ (R)[6a,b]
R^1 = OMe, OCH_2Ph; R^2 = $(CH_2)_2COOMe$[20]
R^1 = COMe; R^2 = $(CH_2)_2COMe$[24]
R^1 = COOEt; R^2 = $(CH_2)_2COOEt$, $CH(COO$-t-Bu$)_2$, $(CH_2)_2COCH_3$ (R)[24]

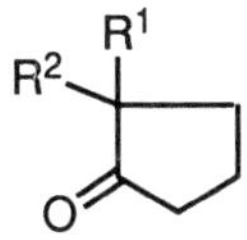

R^1 = CH_2OCH_2OMe
R^2 = $(CH_2)_2COMe$ (S)[6b]
R^1 = Me; R^2 = $(CH_2)_2COOMe$
(R)[6a,b]

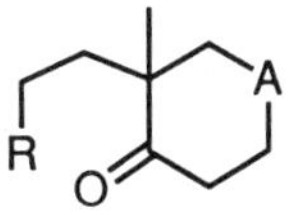

A = N-Me; R = CN (+)[25a]
A = N-Me; R = COOMe (-)[25a]
A = S; R = CO-Me (R and S)[25b]

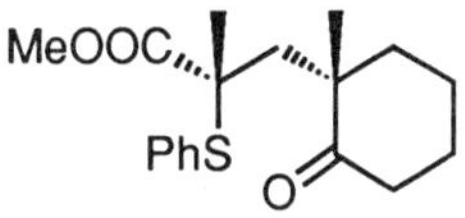

(S,S)[26]

41

11: A = NCH₂Ph; R = Et
(R,R) and (S,S)²⁷ᵃ
12: A = CH₂; R = Me (1S,2R)²⁸

13: A = NCH₂Ph; R = Et (R,R)²⁷ᵃ
14: A = CH₂; R = Me (1S,2R)²⁸

15 : (6S,7R)²⁹

16 : (S)²⁸

17³⁰

Compounds **11-16** arise from an analogous but intramolecular reaction. A special application of the chiral imine procedure, i.e., the use of an oxaziridine rather than an electrophilic olefin, yielding compound **17**, has been reported.

Several of these optically active compounds were used for the synthesis of natural products. Thus (R)- and (S)-dimethyloctalones **5** lead, respectively, to natural and ent-geosmin,[19] while another natural product was obtained from (S,S)-dimethyloctalone **6**, itself a natural compound.[19] Compounds **7** and **8** were used to synthesize a ring C aromatic steroid[21] and a key [ABC] steroid intermediate,[22] respectively.

Tricyclic compound **9** is the starting compound for the preparation of a 14-hydroxyisomorphinan derivative,[23] and (R)-keto ester **10** for the preparation of (+)-cassiol[31] and (-)-19-noraspidospermidine.[32] (R,R)-**11** is an intermediate in the total synthesis of (-)-ajmalicine and (-)-tetrahydroalstonine,[27a] as well as of (+)-yohimbine,[27b] while its (S,S)-enantiomer can led to (-)-(10R)-hydroxydihydroquinine.[27a]

As in the case of methyloctalone **3**, the racemic counterpart of dimethyloctalone **5** has been used for the synthesis of tuberiferine[33] [the (S)-**5** enantiomer would be required for the natural compound] and for the synthesis of isocostic and 3-oxoisocostic acid[34] [(R)-**5** required]. From racemic **6**, both α–vetispirene[35] (S,S) and frullanolide[36] (R,R) have been synthesized.

1. Unité de Chimie Organique (UA CNRS 476), Ecole Supérieure de Physique et Chimie Industrielles de Paris, 10 rue Vauquelin, 75231 Paris Cedex 05, France.
2. "Vogel's Textbook of Practical Organic Chemistry", 5th ed.; Longman Inc.: New York, 1989; p. 143.
3. Toda, F.; Tanaka, K. *Tetrahedron Lett.* **1988**, *29*, 551.
4. Djerassi, C.; Marshall, D. *J. Am. Chem. Soc.* **1958**, *80*, 3986; Adams, W. R.; Chapman, O. L.; Sieja, J. B.; Welstead, W. J. Jr. *J. Am. Chem. Soc.* **1966**, *88*, 162; Touboul, E.; Brienne, M. J.; Jacques, J. *J. Chem. Res. S(Synopsis)* **1977**, 106; Brugidou, J.; Christol, H.; Sales, R. *Bull. Soc. Chim. Fr.* **1979**, *(1-2,Pt. 2)* 40; Marshall, J. A.; Flynn, K. E. *J. Am. Chem. Soc.* **1982**, *104*, 7430; Johnson, C. R.; Zeller, J. R. *J. Am. Chem. Soc.* **1982**, *104*, 4021; Toda, F.; Tanaka, K. *Chem. Lett.* **1985**, 885.
5. Grattan, T. J.; Whitehurst, J. S. *J. Chem. Soc., Chem. Commun.* **1988**, 43; Tanaka, A.; Kamata, H.; Yamashita, K. *Agric. Biol. Chem.* **1988**, *52*, 2043.
6. (a) Pfau, M.; Revial, G.; Guingant, A.; d'Angelo, J. *J. Am. Chem. Soc.* **1985**, *107*, 273; (b) Pfau, M.; Guingant, A.; Revial, G.; d'Angelo, J. PCT Int. Appl. WO 85 04 873, 1985; *Chem. Abst.* **1986**, *105*, 190527j; (c) "Fieser and Fieser's Reagents for Organic Synthesis", Wiley, 1988; Vol. 13, p. 185; Revial G.; Pfau, M. In "Progress in Terpene Chemistry", Frontières Ed.; Paris: France, 1986; p. 383; Sevin, A.; Tortajada, J.; Pfau, M. *J. Org. Chem.* **1986**, *51*, 2671; .Sevin, A.; Masure, D.; Giessner-Prettre, C.; Pfau, M. *Helv. Chim. Acta* **1990**, *73*, 552.

7. Boeckman, R. K., Jr.; Silver, S. M. *J. Org. Chem.* **1975**, *40*, 1755.
8. Dauben, W. G.; Ashcraft, A. C. *J. Am. Chem. Soc.* **1963**, *85*, 3673.
9. Enzell, C. *Tetrahedron Lett.* **1962**, 185.
10. Johnson, C. R.; Barbachyn, M. R. *J. Am. Chem. Soc.* **1982**, *104*, 4290.
11. Cuomo, J. *J. Agric. Food Chem.* **1985**, *33*, 717.
12. Wijnberg, J. B. P. A.; Jongedijk, G.; de Groot, A. *J. Org. Chem.* **1985**, *50*, 2650.
13. Marshall, J. A.; Pike, M. T.; Carroll, R. D. *J. Org. Chem.* **1966**, *31*, 2933.
14. Wijnberg, J. B. P. A.; Vader, J.; de Groot, A. *J. Org. Chem.* **1983**, *48*, 4380.
15. Heathcock, C. H.; Kelly, T. R. *Tetrahedron* **1968**, *24*, 1801.
16. Heathcock, C. H.; Kelly, T. R. *J. Chem. Soc., Chem. Commun.* **1968**, 267.
17. Vig, O. P.; Anand, R. C.; Kumar, B.; Sharma, S. D. *J. Indian Chim. Soc.* **1968**, *45*, 1033.
18. Brown, P. A.; Jenkins, P. R.; Fawcett, J.; Russell, D. R. *J. Chem. Soc., Chem. Commun.* **1984**, 253; Brown, P. A.; Jenkins, P. R. *J. Chem. Soc., Perkin Trans. I* **1986**, 1303.
19. Revial, G. *Tetrahedron Lett.* **1989**, *30*, 4121.
20. Desmaële, D.; d'Angelo, J. *Tetrahedron Lett.* **1989**, *30*, 345.
21. Volpe, T.; Revial, G.; Pfau, M.; d'Angelo, J. *Tetrahedron Lett.* **1987**, *28*, 2367.
22. d'Angelo, J.; Revial, G.; Volpe, T.; Pfau, M. *Tetrahedron Lett.* **1988**, *29*, 4427.
23. Sdassi, H.; Revial, G.; Pfau, M.; d'Angelo, J. *Tetrahedron Lett.* **1990**, *31*, 875.
24. Brunner, H.; Kraus, J.; Lautenschlager, H.-J. *Monatsh. Chem.* **1988**, *119*, 1161.
25. (a) Gaidarova, E. L.; Grishina, G. V.; Potapov, V. M.; Kornilov, M. Yu.; Kozhinskaya, M. V. U.S.S.R. SU Patent 1 384 579, 1988; *Chem. Abstr.* **1988**, *109*, 190258e; (b) Matsuyama, H.; Ebisawa, Y.; Kobayashi, M.; Kamigata, N. *Heterocycles* **1989**, *29*, 449.
26. d'Angelo, J.; Guingant, A.; Riche, C.; Chiaroni, A. *Tetrahedron Lett.* **1988**, *29*, 2667.

27. (a) Hirai, Y.; Terada, T.; Yamazaki, T. *J. Am. Chem. Soc.* **1988**, *110*, 958; (b) Hirai, Y.; Terada, T.; Okaji, Y.; Yamazaki, T.; Momose, T. *Tetrahedron Lett.* **1990**, *31*, 4755.
28. Dumas, F.; d'Angelo, J. *Tetrahedron: Asymmetry* **1990**, *1*, 167.
29. d'Angelo, J.; Ferroud, C.; Riche, C.; Chiaroni, A. *Tetrahedron Lett.* **1989**, *30*, 6511.
30. Davis, F. A.; Sheppard, A. C. *Tetrahedron Lett.* **1988**, *29*, 4365.
31. Takemoto, T.; Fukaya, C.; Yokoyama, K. *Tetrahedron Lett.* **1989**, *30*, 723.
32. d'Angelo, J.; Desmaele, D. *Tetrahedron Lett.* **1990**, *31*, 879; Desmaele, D.; d'Angelo, J. *Tetrahedron Lett.* **1990**, *31*, 883.
33. Grieco, P. A.; Nishizawa, M. *J. Chem. Soc., Chem. Commun.* **1976**, 582.
34. Cruz, R.; Martinez, R. M. *Aust. J. Chem.* **1982**, *35*, 451.
35. Caine, D.; Boucugnani, A. A.; Chao, S. T.; Dawson, J. B.; Ingwalson, P. F. *J. Org. Chem.* **1976**, *41*, 1539.
36. Still, W. C.; Schneider, M. J. *J. Am. Chem. Soc.* **1977**, *99*, 948.

Appendix

Chemical Abstracts Nomenclature (Collective Index Number); (Registry Number)

(R)-(-)-10-Methyl-1(9)-octal-2-one: 2(3H)-Naphthalenone, 4,4a,5,6,7,8-hexahydro-4a-methyl-, (R)- (10); (63975-59-7);

Imines 1: Benzenemethanamine, α–methyl-N-(2-methylcyclohexylidene)- (10); (76947-33-6)

(S)-(-)-α–Methylbenzylamine: Benzylamine, α–methyl-, (S)-(-)- (8); Benzenemethanamine, α–methyl-, (S)- (9); (2627-86-3)

2-Methylcyclohexanone: Cyclohexanone, 2-methyl (8,9); (583-60-8)

(R)-(+)-2-Methyl-2-(3-oxobutyl)cyclohexanone: Cyclohexanone, 2-methyl-2-(3-oxobutyl)-, (R)- (11); (91306-30-8)

Methyl vinyl ketone: 3-Buten-2-one (8,9); (78-94-4)

Imine of compound 2: 2-Butanone, 4-[1-methyl-2-[(1-phenylethyl)imino]cyclohexyl]-, [S-(R*,S*)]- (11); (94089-44-8)

(R,R)-1,2-DIPHENYL-1,2-ETHANEDIOL (STILBENE DIOL)

(1,2-Ethanediol, 1,2-diphenyl-[R-(R*,R*)]-)

R' = p-chlorobenzoyl

Submitted by Blaine H. McKee,[1] Declan G. Gilheany,[2] and K. Barry Sharpless.[1]
Checked by Aaron Balog and Robert K. Boeckman, Jr.

1. Procedure

To a 3-L, three-necked, round-bottomed flask equipped with a mechanical stirrer and two glass stoppers at room temperature are added (E)-1,2-diphenylethene (trans-stilbene) (180.25 g, 1.0 mol, 1.0 equiv) (Note 1), 4-methylmorpholine N-oxide (NMO) [260 mL of a 60% by wt. aqueous solution (1.5 mol, 1.5 equiv) Notes 1 and 2], dihydroquinidine 4-chlorobenzoate (23.25 g, 0.05 mol, 0.05 equiv) (Notes 3 and 4), 375 mL of acetone and 7.5 mL water. [The solution is 0.1 M in alkaloid (Note 5), 2 M in olefin, and the solvent is 25% water/75% acetone (v/v) (Note 6)]. The flask is immersed in a 0°C cooling bath and stirred for 1 hr. Osmium tetroxide (1.0 g, 4.0 mmol, 4.0 x 10^{-3} equiv) is added in one portion, producing a milky brown-yellow suspension (Note 7). The reaction mixture is then stirred at 0°C for 33 hr and monitored by silica TLC (3:1 $CH_2Cl_2:Et_2O$ v/v) until complete. At this point, the mixture is warmed to room temperature, diluted with 500 mL of dichloromethane, and sodium metabisulfite (285 g, 1.5 mol) is added in several portions while the internal temperature is maintained at room temperature with an ice bath as needed. After addition is complete and the exothermic reaction has subsided, stirring is continued at room temperature for 1 hr (Note 8). Anhydrous sodium sulfate (50 g) is added and the mixture is stirred at room temperature overnight (Note 9). The suspension is filtered through a 20-cm Büchner funnel, the filtrand is rinsed thoroughly with acetone (3 x 250 mL), and the filtrate is concentrated to a brown paste (Note 10). The paste is dissolved in 3.5 L of ethyl acetate, transferred to a 6-L separatory funnel, and washed sequentially with water (2 x 500 mL) (Note 11), 0.25 M sulfuric acid (2 x 500 mL) (Note 12), and brine (1 x 500 mL). The initial, aqueous washes are kept separate from the subsequent acid washes which are retained for alkaloid recovery (Notes 13 and 14). The organic layer is dried (Na_2SO_4), and concentrated to give the crude diol in quantitative yield (222.7 g, 1.04 mol, 104%). The ee of the crude product is determined by ^1H NMR analysis of the derived bis-Mosher ester to be 90%. One recrystallization[3] from hot aqueous 95% ethanol (3 mL/g) affords 155-162 g (72-75%) of enantiomerically pure stilbene diol as a white solid, mp 144.5-146.5°C, $[\alpha]_D^{25}$ 90.0° (abs EtOH, c 1.96) (Note 15).

2. Notes

1. trans-Stilbene and N-methylmorpholine N-oxide were obtained from the Aldrich Chemical Company, Inc.

2. This solution contains 173.7 g of NMO and 117.5 mL of water. Its density is 1.130 g/mL.

3. Dihydroquinidine 4-chlorobenzoate is available from the Aldrich Chemical Company, Inc. The optical rotation of the commercial sample employed by the checkers had an $[\alpha]_D^{25}$ of -68.9° (EtOH, c 0.95).

4. Many other dihydroquinidine derivatives have been assayed in the catalytic, asymmetric dihydroxylation reaction (ADH)[4] and the submitters have recently found that the benzoate and 2-naphthoate esters are slightly better for aryl-substituted alkenes while certain ethers are better for other substrates.[5] However, since the level of asymmetric induction is already high, there is little advantage to be gained from their use in this case.

5. When the alkaloid concentration is increased to 0.25 M there is a slight increase in ee; when the concentration is decreased below 0.067 M there is a drastic decrease in ee.[6]

6. These solvent conditions have been optimized. The low solubility of stilbene in the reaction mixture approximates "slow addition" conditions.[7]

7. Osmium tetroxide is volatile and toxic and therefore should be used only in a well-ventilated hood. On a 1-mole scale, osmium tetroxide was added as a solid. On a smaller scale, it was added as a solution (ca. 0.5 M) in toluene.[6a]

8. The checkers noted a significant exotherm upon addition of the sodium metabisulfite and warming to room temperature that caused the temperature of the methylene chloride solution to rise to the boiling point. Addition of the bisulfite in small portions at 0°C had no beneficial effect in moderating the exotherm that occurred after warming to room temperature.

9. This time can be reduced to 30 min without any deleterious effects.[7]

10. The filtrate is concentrated on a rotary evaporator with slight heating (bath temperature 30-40°C). In some runs with other substrates, stronger heating (bath temperature 70-80°C) caused the reaction mixture to turn black. However, there was no significant effect on either yield or enantiomeric excess in those cases.

11. These washes remove 4-methylmorpholine as well as any remaining acetone. Subsequent contact of the diol with acetone should be avoided to prevent any chance of acetonide formation.

12. It is important to use sulfuric acid at this point to ensure efficient extraction. The sulfate salt of the alkaloid is more soluble in water and less soluble in organic solvents than the hydrochloride salt. In the ADH of other alkenes the preferred system is sulfuric acid/diethyl ether. However stilbene diol is only sparingly soluble in diethyl ether, which necessitates the use of ethyl acetate. Chlorinated hydrocarbon solvents should be avoided since both alkaloid salts have appreciable solubility in them. When diethyl ether is used as the organic phase, not all of the reaction mixture dissolves in it, but the material that remains undissolved is derived solely from 4-methylmorpholine.

13. Back extraction of the acid layer yields an insignificant amount of diol in this case. However, it may be necessary for more water-soluble diols. For example, in the case of the diol from methyl 2-octenoate the yield is increased by 30% with one back extraction, while styrene glycol requires repeated prolonged extraction.

14. The alkaloid was recovered by raising the pH of the acidic washes to 11 with sodium carbonate, transferring the solution to a 6-L separatory funnel and extracting with methylene chloride (3 x 500 mL). The alkaloid was recovered in 84-85% yield as a white foam and was used without further purification in subsequent dihydroxylations. The use of recovered alkaloid by the checkers resulted in a decrease in the ee of the crude diol to 80%. The submitters, however, report ee's of 90% from repeated use of recovered alkaloid. Note that the alkaloid is stable for several days in the acidic aqueous extract. Once the solution is made alkaline, however, it should be extracted immediately.

15. Optically pure S,S-stilbene diol can be similarly obtained in 66% yield using dihydroquinine-4-chlorobenzoate. The crude ee before recrystallization is 74%.

Waste Disposal Information

All toxic materials were disposed of in accordance with "Prudent Practices for Disposal of Chemicals from Laboratories"; National Academy Press; Washington, DC, 1983.

3. Discussion

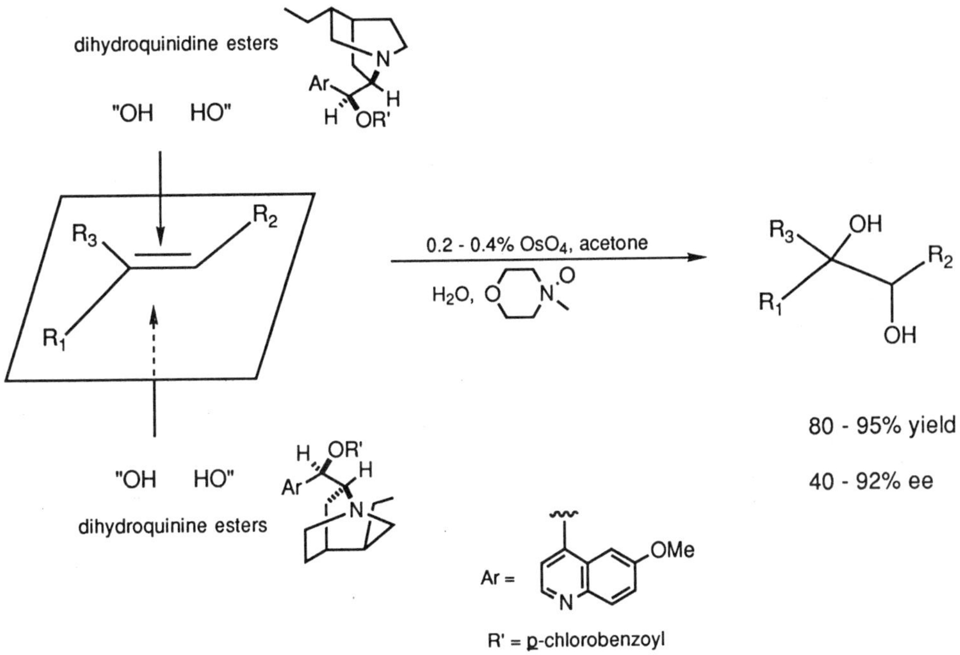

80 - 95% yield
40 - 92% ee

Ar = 6-methoxyquinolin-4-yl

R' = p-chlorobenzoyl

The present procedure describes a convenient preparation of threo-stilbene diol on a 1-mole scale and illustrates the utility of the catalytic, asymmetric dihydroxylation (ADH) of solid substrates on a large scale. Note that this procedure calls for more water than initially reported.[6a] The extra water leads to higher ee (90% cf. 78%[6a]) by better approximating the slow addition conditions that are now almost always used for liquid olefins.[7,8]

Recently reported uses of optically pure stilbene diol in asymmetric synthesis include. (1) the dimethyl ether as a ligand for effecting enantioselective conjugate addition;[9] (2) the preparation of α,β-unsaturated ketals for achieving diastereoselective Simmons-Smith cyclopropanation;[10] (3) the preparation of enantiomercially pure β-halohydrins;[11] and (4) the preparation of chiral crown ethers.[12]

1. Department of Chemistry, Massachusetts Institute of Technology, Cambridge, MA 02139.
2. On leave from the Department of Chemistry, Saint Patrick's College, Maynooth, Ireland. D.G.G. acknowledges the receipt of a grant from the Irish Scholarship Exchange (Fulbright) program.
3. Jacques, J.; Collet, A.; Wilen, S. H. in "Enantiomers, Racemates and Resolutions"; Wiley Interscience: New York, 1981; pp. 224-225.
4. Blackburn, B. K.; Gilheany, D. G.; Pearlstein, R. M.; Markó, I.; Mungall, W. S.; Schröder, G.; Sharpless, K. B., unpublished results.
5. Shibata, T.; Gilheany, D. G.; Blackburn, B. K.; Sharpless, K. B. *Tetrahedron Lett.* **1990**, *31*, 3817.
6. (a) Jacobsen, E. N.; Markó, I.; Mungall, W. S.; Schröder, G.; Sharpless, K. B. *J. Am. Chem. Soc.* **1988**, *110*, 1968; (b) McKee, B. H.; Sharpless, K. B., unpublished results.
7. Wai, J. S. M.; Markó, I.; Svendsen, J. S.; Finn, M. G.; Jacobsen, E. N.; Sharpless, K. B. *J. Am. Chem. Soc.* **1989**, *111*, 1123.

8. Lohray, B. B.; Kalantar, T. H.; Kim, B. M.; Park, C. Y.; Shibata, T.; Wai, J. S. M.; Sharpless, K. B. *Tetrahedron Lett.* **1989**, *30*, 2041.
9. Tomioka, K.; Nakajima, M.; Koga, K. *J. Am. Chem. Soc.* **1987**, *109*, 6213.
10. Mash, E. A.; Torok, D. S. *J. Org. Chem.* **1989**, *54*, 250.
11. Konopelski, J. P.; Boehler, M. A.; Tarasow, T. M. *J. Org. Chem.* **1989**, *54*, 4966.
12. Crosby, J.; Fakley, M. E.; Gemmell, C.; Martin, K.; Quick, A.; Slawin, A. M. Z.; Shahriari-Zavareh, H.; Stoddart, J. F.; Williams, D. J. *Tetrahedron Lett.* **1989**, *30*, 3849.

Appendix
Chemical Abstracts Nomenclature (Collective Index Number); (Registry Number)

(R,R)-1,2-Diphenyl-1,2-ethanediol: 1,2-Ethanediol, 1,2-diphenyl-, [R-(R*,R*)]- (9); (52340-78-0)

(E)-1,2-Diphenylethene: Stilbene, (E)- (8); Benzene, 1,1'-(1,2-ethenediyl)bis-, (E)- (9); (103-30-0)

4-Methylmorpholine N-oxide: Morpholine, 4-methyl-, 4-oxide (8,9); 7529-22-8)

Dihydroquinidine 4-chlorobenzoate: Cinchonan-9-ol, 10,11-dihydro-6'-methoxy-, 4-chlorobenzoate (ester), (9S)- (12); (113162-02-0)

Osmium tetroxide: Osmium oxide (8); Osmium oxide, (T-4) (9); (20816-12-0)

A STABLE CHIRAL 1,4-DIHYDROPYRIDINE EQUIVALENT FOR THE ASYMMETRIC SYNTHESIS OF SUBSTITUTED PIPERIDINES: 2-CYANO-6-PHENYLOXAZOLOPIPERIDINE

(5H-Oxazolo[3,2-a]pyridine-5-carbonitrile, hexahydro-3-phenyl-, [3R-(3α,5β,8aβ])

Submitted by Martine Bonin, David S. Grierson, Jacques Royer, and Henri-Philippe Husson.[1]

Checked by Gilbert Rishton and Larry E. Overman.

1. Procedure

CAUTION! Aqueous potassium cyanide is used in this procedure. All operations should be conducted in a well-ventilated hood and rubber gloves should be worn.

2-Cyano-6-phenyloxazolopiperidine. A 2-L, round-bottomed flask is charged with 10 g (0.073 mol) of (-)-phenylglycinol (Note 1) and 40 g of citric acid in 1 L of distilled water. The mixture is stirred magnetically until complete dissolution is achieved and is cooled to 0-5°C (ice-water bath). The flask is equipped with a dropping funnel and 45 mL of an aqueous 24% glutaraldehyde solution (0.11 mol) is added dropwise over 20 min and the resultant cloudy solution is stirred for an additional 30 min at 0°C. The cooling bath is removed and a solution of 7.15 g (0.11 mol) of potassium cyanide in 20 mL of water and 200 mL of methylene chloride are added sequentially. The resulting two-phase reaction system is stirred for 3 hr at room temperature, then the aqueous phase is neutralized (Note 2) by addition of sodium bicarbonate and the two layers are separated. The water layer is extracted with three, 200-mL portions of methylene chloride (Note 3) and the combined methylene chloride layers are dried over sodium sulfate and concentrated on a rotary evaporator to a volume of 500 mL. Zinc bromide (2 g) is added in small portions over 5 min to this solution and vigorous stirring under nitrogen is continued for 3 hr (Note 4). The reaction mixture is concentrated to a volume of approximately 150 mL (Note 5) and the resultant mixture is applied to a 10-cm diameter flash chromatography column prepared using hexane-ether (2:1) as the eluant. The desired product is eluted first (Notes 6, 7). By recrystallization from hexane, the product [10.8-11.6 g (65-70%)] is obtained analytically pure; mp 79-81°C, $[\alpha]_D^{23}$ -280° (CHCl$_3$, c 1.0) (Note 8).

2. Notes

1. The checkers used (R)-(-)-2-phenylglycinol [98%, $[\alpha]_D^{24}$ -31.7° (1 N HCl, c 0.76)] purchased from Aldrich Chemical Company, Inc. The submitters employed material [mp 78°C, $[\alpha]_D^{20}$ -26.5° (MeOH, c 0.7)] prepared by lithium aluminum hydride reduction of (-)-phenylglycine and report that the yield of 2-cyano-6-phenyloxazolopiperidine prepared from this material is 75-83%.

2. Extraction at pH greater than 9 led to the formation of by-products (2,6-dicyanopiperidines)[2] and consequently to a lower yield of the desired product.

3. The aqueous layer containing residual potassium cyanide is destroyed by addition of potassium permanganate.

4. The reaction must be carried out in a well-ventilated hood as hydrogen cyanide may be evolved.

5. Complete evaporation of the solvent gives a viscous oil which is only slightly soluble in either the elution solvent or methylene chloride. Significant loss of material thus occurs during the purification process.

6. R_f = 0.6 (SiO_2, hexane-ether : 2-1) for the product.

7. Further elution permits the isolation of a mixture of two other isomers (0.3 g, 3.6%).

8. The product obtained by this procedure shows the following spectral data: IR ($CHCl_3$) cm^{-1}: 2100; ^{1}H NMR (400 MHz, $CDCl_3$) δ: 1.5-2.0 (m, 5 H); 2.13 (dd, 1 H, J = 11.5, J' = 1.5); 3.74 (t, 1 H, J = 7.8); 3.85 (bd, 1 H, J = 7.1); 3.90 (t, 1 H, J = 8.0); 4.12 (dd, 1 H, J = 9.7, J' = 2.8); 4.25 (t, 1 H, J = 7.9); 7.4 (m, 5 H); ^{13}C NMR ($CDCl_3$, 15 MHz) δ: 19.3; 28.0; 30.0; 47.4; 63.9; 73.0; 89.9; 116, 128.2, 128.6; 129.0; 137.4.

Waste Disposal Information

All toxic materials were disposed of in accordance with "Prudent Practices for Disposal of Chemicals from Laboratories"; National Academy Press; Washington, DC, 1983.

3. Discussion

The preparation described here is an improvement of the previous procedure.[3] The double condensation of glutaraldehyde with the amino group of (R)-(-)-phenylglycinol (related to the Robinson-Schopf condensation) probably leads to the expected product via the formation of an intermediate of type A[4] and/or B.

Trapping intermediate B with cyanide ion would lead to the final product. Formation of the four possible product isomers has been observed after short reaction periods, and equilibration of this mixture to the described product **2** has been demonstrated. In the original procedure compound **2** was obtained in 50-60% yield after prolonged (72 hr) reaction of the reactants in water. In the present procedure equilibration of all intermediates and/or product isomers to the observed, thermodynamically more stable form of compound **2** is accelerated and the yield improved considerably through the use of zinc bromide as a catalyst in an organic medium.

2-Cyano-6-phenyloxazolopiperidine, **2**, is a stable, chiral, 1,4-dihydropyridine equivalent, useful for the asymmetric synthesis of piperidines. When synthon **2** was used, the asymmetric syntheses of several alkaloids or analogs have been achieved and reported in the literature. A sequence of reactions involving alkylation at the α-aminonitrile center (LDA, RX), elimination of the cyano group with concomitant opening of the oxazolidine ring (NaBH$_4$), and finally debenzylation (H$_2$, Pd/C) permits the preparation of α-alkylated piperidine **3** in both high yield (90% overall) and ee (≥ 95%).[5] If one starts from the same compound **2**, enantiomeric **4** can also be prepared.[5] A regio- and chemoselective elimination of the cyano group is also possible giving an oxazolopiperidine intermediate which can further be used to prepare the cis- and trans-2,6-disubstituted piperidines **5**[5] and **6**.[6] When the appropriate electrophile is used, enantioselective syntheses of indolizidines **7**,[3] **8**,[7] decahydroquinolines **9**,[8] and amino alcohol **10**[9] have been achieved.

The asymmetric syntheses of β–amino alcohols have also been reported recently in the literature[10,11] using chiral compound **2** as a starting material.

Scheme

1. Institut de Chimie des Substances Naturelles du C.N.R.S., 91198 Gif-sur-Yvette Cedex, France.
2. Bonin, M.; Chiaroni, A.; Riche, C.; Beloeil, J.-C.; Grierson, D. S.; Husson, H.-P. *J. Org. Chem.* **1987**, *52*, 382.
3. Royer, J.; Husson, H.-P. *J. Org. Chem.* **1985**, *50*, 670.

4. Quirion, J.-C.; Grierson, D. S.; Royer, J.; Husson, H.-P. *Tetrahedron Lett.* **1988**, *29*, 3311.
5. Guerrier, L.; Royer, J.; Grierson, D. S.; Husson, H.-P. *J. Am. Chem. Soc.* **1983**, *105*, 7754.
6. Grierson, D. S.; Royer, J.; Guerrier, L.; Husson, H.-P. *J. Org. Chem.* **1986**, *51*, 4475.
7. Royer, J.; Husson, H.-P. *Tetrahedron Lett.* **1985**, *26*, 1515.
8. Bonin, M.; Royer, J.; Grierson, D. S.; Husson, H.-P. *Tetrahedron Lett.* **1986**, *27*, 1569.
9. Ratovelomanana, V.; Royer, J.; Husson, H.-P. *Tetrahedron Lett.* **1985**, *26*, 3803.
10. Delgado, A.; Mauleon, D. *Synth. Commun.* **1988**, *18*, 823.
11. McIntosh, J. M.; Matassa, L. C. *J. Org. Chem.* **1988**, *53*, 4452.

Appendix
Chemical Abstracts Nomenclature (Collective Index Number); (Registry Number)

2-Cyano-6-phenyloxazolopiperidine: 5H-Oxazolo[3,2-a]pyridine-5-carbonitrile, hexahydro-3-phenyl-, [3R-(3α,5β,8aβ)]- (11); (88056-92-2)

Potassium cyanide (8,9); (151-50-8)

(-)-Phenylglycinol: Benzeneethanol, β–amino-, (R)- (9); (56613-80-0)

Citric acid (8); 1,2,3-Propanetricarboxylic acid, 2-hydroxy- (9); (77-92-9)

Glutaraldehyde (8); Pentanedial (90: (111-30-8)

Zinc bromide (8,9); (7699-45-8)

(S)-(-)- AND (R)-(+)-1,1'-BI-2-NAPHTHOL
([1,1'-Binaphthalene]-2,2'-diol, (S)- and [1,1'-Binaphthalene]-2,2'-diol, (R)-)

Submitted by Romas J. Kazlauskas.[1]

Checked by Mark R. Sivik and Leo A. Paquette.

1. Procedure

A. Racemic 1,1'-bi-2-naphthyl pentanoate. A suspension of 203 g (0.71 mol) of racemic 1,1'-bi-2-naphthol (Note 1) and 215 mL (1.54 mol) of triethylamine in 2 L of ethyl ether is stirred magnetically in a 4-L Erlenmeyer flask. Over a period of 20 min 185 mL (1.56 mol) of pentanoyl chloride is added (Note 2) and the suspension is stirred for an additional hour to ensure complete reaction (Note 3). The mixture is poured into a 6-L separatory funnel and washed twice with 2-L portions of aqueous 1 M sodium bicarbonate and once with a 2-L portion of water resulting in a clear yellow-orange ether solution.

B. (S)-(-)-1,1'-Bi-2-naphthol. In a 12-L, round-bottomed flask the above ether solution is diluted to 4 L with additional ethyl ether. A 5.0-mL aliquot of this solution is used to assay the enzyme (Note 4). The remaining ether solution is stirred using an overhead stirrer with 4 L of aqueous 0.1 M phosphate buffer (pH 7.5) containing 60 g of crude sodium taurocholate (Note 5). An opaque emulsion forms. The reaction is started by adding 2000 units of cholesterol esterase activity (100-150 g of bovine pancreas acetone powder, Note 4). Stirring is continued at ~25°C and the flask is stoppered to minimize evaporation of ether. The pH of the emulsion is measured occasionally and readjusted to 7.2 ± 0.3 by adding aqueous 1 M sodium hydroxide. Approximately 250 mL of base is consumed during the first 3 hr; an additional 400 mL is consumed over the next 20 hr. Although the consumption of base virtually ceases after 24 hr stirring is continued for a total of 3 days (Note 6).

To break the emulsion 400 mL of ethanol is added and the mixture is transferred to separatory funnels and allowed to settle for 4 hr. Three layers form: at the top - a clear yellow ether phase, at the bottom - a brown aqueous phase and in between - an opaque emulsion layer. The brown aqueous layer is discarded. The emulsion layer (~1 L) is transferred to a flask and broken up by addition of 200 g of magnesium sulfate in portions. Heat is evolved, the ether boils and two layers form. This ether layer is combined with the first ether layer, dried over ~50 g of magnesium sulfate, filtered and concentrated by rotary evaporation to ~300 mL of an orange oil. Toluene (500 mL) is added and the solution cooled to 4°C overnight. The fine white crystals (70-71 g) are collected by filtration and washed twice with 20-mL portions of cold toluene (Note 7). The filtrate is concentrated to ~600 mL by rotary evaporation and cooled once again to 4°C. The additional 10 g of crystals which form are collected by filtration and washed twice with 20-mL portions of toluene. Recrystallization of the combined crystals from 375-400 mL of toluene yields 65-68 g (64-67%) of white crystals, mp 211-213.5°C; $[\alpha]_D^{19}$ -33.2 ± 0.8° (THF, c 0.2); >99% diol (see Note 3), >99.9% enantiomeric purity (Note 8).

C. *(R)-(+)-1,1'-Bi-2-naphthol.* Toluene in the above filtrate is removed by rotary evaporation, and the residue is recrystallized from methanol (500 ml) overnight at 4°C. Yellow crystals (125-141 g) form and are collected by filtration and washed twice with 20 mL of hexane. Recrystallization from 500 mL of methanol yields pure (R)-binaphthol dipentanoate [mp 63-65°C, 89-102 g, >99% ee (R), $[\alpha]_D^{19}$ +15.0 ± 0.3° (CHCl$_3$, c 0.4), Note 9]. If desired, an additional 25 g of dipentanoate can be isolated from the filtrate by column chromatography on 1 kg of silica gel eluted with methylene chloride followed by crystallization (Note 9).

The crystalline dipentanoate (89-102 g, 0.20 mol) is dissolved in 1 L of methanol containing 6.6 g (0.12 mol) of sodium methoxide. After 4 hr at room temperature, analysis of the solution by thin layer chromatography (Note 3) shows only traces of the mono- and diester. The solution is neutralized to pH <7 (test paper) with ~10 mL of concd hydrochloric acid. The solution is diluted with 1 L of 0.1 M phosphate buffer (pH 7), transferred to a 4-L separatory funnel and extracted with a mixture of 1 L of ethyl ether and 500 mL of toluene. The organic layer is washed with a 1-L portion of water, dried over magnesium sulfate, concentrated to 300 mL and cooled to 4°C.

White crystals (48-64 g) separate and are collected by filtration and washed twice with 20-mL portions of cold toluene, mp 211-213.5°C; $[\alpha]_D^{19}$ +33.9 ± 0.2° (THF, c 0.2); 99% chemical purity (Note 3), >99% enantiomeric purity (Note 8).

2. Notes

1. (±)-1,1'-Bi-2-naphthol was purchased from Aldrich Chemical Company, Inc. or prepared by oxidative coupling of 2-naphthol.[2]

2. The initial suspension is thick and can sometimes be difficult to stir magnetically. In this case, occasional swirling by hand is sufficient. The mixture thins as the reaction proceeds. *Caution: This exothermic reaction causes the ether to boil; pentanoyl chloride should be added slowly, allowing the heat of reaction to dissipate.* The checkers cooled the reaction mixture in an ice bath during addition of the acid chloride. Pentanoyl chloride was obtained from Aldrich Chemical Company, Inc.

3. To ensure high enantiomeric purity of the product there should be <0.5% 1,1'-bi-2-naphthol or its monoester in this solution. The relative amounts of binaphthol species can be accurately determined by HPLC on a reverse-phase column eluted with a water-acetonitrile gradient (50-100% over 10 min). Both 1,1'-bi-2-naphthol and its dipentanoate have equal (within 2%) extinction coefficients at 254 nm. The monopentanoate absorbs more strongly: the relative extinction coefficient at 254 nm is 1.13. Alternatively, the solution composition can be estimated using thin layer chromatography: silica gel eluted with 1:4 ethyl acetate/cyclohexane: 1,1'-bi-2-naphthol, R_f 0.39; monopentanoate, R_f 0.56; dipentanoate, R_f 0.71.

4. The catalyst for this reaction is the enzyme cholesterol esterase (EC 3.1.1.13). Bovine pancreas acetone powder (Sigma Chemical Company), a crude extract from pancreas, is an inexpensive source of cholesterol esterase activity. This extract contains ~15 units of cholesterol esterase activity/gram; unit = μmol of ester

hydrolyzed/min. To measure the activity, the ethereal aliquot of binaphthol dipentanoate is stirred rapidly using a magnetic stirrer with 5.0 mL of 10 mM phosphate buffer (pH 7.0) containing 75 mg of crude sodium taurocholate (Sigma Chemical Company). Approximately 200 mg of acetone powder is added and the pH of the emulsion is monitored with a pH meter and maintained at 7.0 by addition of aqueous 0.1 M sodium hydroxide in portions of 50 µL until ~200 µL has been added, ~70 min. The slope of a plot of µmoles of base consumed vs. time gives the activity of the acetone powder. (The amount of base needed to readjust the pH to 7.0 after the addition of the slightly acidic acetone powder is ignored in the activity calculation.)

5. Directions for the preparation of this buffer solution are given in reference 3.

6. After 24 hr, analysis by HPLC shows 37% binaphthol, 10% monopentanoate and 53% dipentanoate, after an additional 2 days of stirring, analysis shows 45%, 3%, and 52%. Isolation of binaphthol and diester by crystallization is substantially more difficult and less efficient from reaction mixtures containing <40 mol % binaphthol.

7. Binaphthol may not crystallize if the solution is wet. If no crystals form, water can be removed by rotary evaporation of the water-toluene azeotrope.

8. Enantiomeric purity of binaphthol is determined using chiral stationary phase HPLC: Pirkle Type 1-A column (Regis Chemical Company) eluted with 20:1 hexane/2-propanol[4] or poly(triphenylmethyl)methacrylate on silica gel (Chiralpak OT, Daicel Chemical Industries, LTD) eluted with methanol.[5] To determine enantiomeric purities >99% ee an HPLC trace of the unknown is compared to the HPLC trace of unknown containing 0.2% deliberately-added racemic material.

9. Crystallization of (R)-binaphthyl dipentanoate increases its enantiomeric purity from ~92% ee in the reaction mixture to >99% ee. The enantiomeric purity of the final product, binaphthol, is not increased by crystallization. The recrystallization step for the dipentanoate ensures high enantiomeric purity. Usually crystallization from methanol must be induced by scratching the side of the flask with a glass rod.

The enantiomeric purity of the dipentanoate is determined after cleavage to binaphthol. A sample of dipentanoate is treated with an equivalent of sodium methoxide in methanol. After 30 min the solution is neutralized with excess acetic acid and analyzed by HPLC as in Note 8.

Waste Disposal Information

All toxic materials were disposed of in accordance with "Prudent Practices for Disposal of Chemicals from Laboratories"; National Academy Press; Washington, DC, 1983.

3. Discussion

Enantiomerically pure binaphthol is used as a chiral auxiliary.[6] For example, it has been used to prepare chiral aluminum hydride reducing agents,[7] chiral Lewis acids catalysts,[8] and chiral crown ethers.[9]

The best previous resolution of binaphthol uses fractional crystallization of the diastereomeric cinchonine salts of binaphthol cyclic phosphate ester.[10] The resolution using cholesterol esterase involves fewer manipulations and thus is simpler and faster than the cinchonine method. Fewer manipulations also enable the resolutions using cholesterol esterase to be carried out on a larger scale. The high enantioselectivity of cholesterol esterase assures high ee for the (S)-enantiomer, while crystallization of (R)-binaphthyl dipentanoate assures high enantiomeric purity for the (R)-enantiomer.

Octahydrobinaphthol and several spirobiindanols can also be resolved using this method, but several bromo-substituted binaphthols could not be resolved because their esters were not hydrolyzed.[11]

1. Department of Chemistry, McGill University, 801 Sherbrooke St., W., Montreal, PQ H3A 2K6 Canada. Initial work was done at General Electric Company, Corporate Research and Development, Schenectady, NY.
2. Pummerer, R.; Prell, E.; Reiche, A. *Ber.* **1926**, *59B*, 2159; note correction in Note 12 of Reiche, A.; Jungholt, K.; Frühwald, E. *Ber.* **1931**, *64B*, 578.
3. Irochijima, S.; Kojima, N. *Agric. Biol. Chem.* **1980**, *46*, 1593.
4. Pirkle, W. H.; Schreiner, J. L. *J. Org. Chem.* **1981**, *46*, 4988.
5. Okamoto, Y.; Honda, S.; Okamoto, I.; Yuki, H. *J. Am. Chem. Soc.* **1981**, *103*, 6971.
6. Review (in Japanese): Miyano, S.; Hashimoto, H. *Yuki Gosei Kagaku Kyokaishi* **1986**, *44*, 713; *Chem. Abstr.* **106**, 137988e.
7. Noyori, R.; Tomino, I.; Tanimoto, Y.; Nishizawa, M. *J. Am. Chem. Soc.* **1984**, *106*, 6709; Noyori, R.; Tomino, I.; Yamada,. M.; Nishizawa, M. *J. Am. Chem. Soc.* **1984**, *106*, 6717.
8. Sakane, S.; Maruoka, K.; Yamamoto, H. *Tetrahedron Lett.* **1985**, *26*, 5535; Maruoka, K.; Itoh, T.; Shirasaka, T.; Yamamoto, H. *J. Am. Chem. Soc.* **1988**, *110*, 310.
9. Sogah, G. D. Y.; Cram, D. J. *J. Am. Chem. Soc.* **1979**, *101*, 3035.
10. Jacques, J.; Fouquey, C. *Org. Synth.* **1989**, *67*, 1; Truesdale, L. K. *Org. Synth.* **1989**, *67*, 13; Kyba, E. P.; Gokel, G. W.; de Jong, F.; Koga, K.; Sousa, L. R.; Siegel, M. G.; Kaplan, L.; Sogah, G. D. Y.; Cram, D. J. *J. Org. Chem.* **1977**, *42*, 4173.
11. Kazlauskas, R. J. *J. Am. Chem. Soc.* **1989**, *111*, 4953.

Appendix

Chemical Abstracts Nomenclature (Collective Index Number);

(Registry Number)

(S)-(-)-1,1'-Bi-2-naphthol: [1,1'-Binaphthalene]-2,2'-diol, (S)-(-)- (8);

[1,1'-Binaphthalene]-2,2'-diol, (S)- (9); (18531-99-2)

(R)-(+)-1,1'-Bi-2-naphthol: [1,1'-Binaphthalene]-2,2'-diol, (R)-(+)- (8);

[1,1'-Binaphthalene]-2,2'-diol, (R)- (9); (18531-94-7)

(±)-1,1'-Bi-2-naphthyl pentanoate: Pentanoic acid, [1,1'-binaphthalene]-2,2'-diyl ester, (±)- (12); (100465-51-8)

(±)-Bi-2-naphthol: [1,1'-Binaphthalene]-2,2'-diol, (±)- (9); (41024-90-2)

Pentanoyl chloride (8,9); (638-29-9)

(R)-1,1'-Bi-2-naphthyl pentanote: Pentanoic acid, [1,1'-binaphthalene]-2,2'-diyl ester, (R)- (12); (110902-38-0)

3,4-DIETHYLPYRROLE AND 2,3,7,8,12,13,17,18-OCTAETHYLPORPHYRIN

(21H,23H-Porphine, 2,3,7,8,12,13,17,18-octaethyl-)

A. O=CH-CH₂-CH₃ + O₂N-CH₂-CH₂-CH₃ →[KF/IPA] 1 (3-nitro-4-hydroxyhexane)

B. **1** + Ac₂O →[H₂SO₄] **2** (3-nitro-4-acetoxyhexane)

C. **2** + C=NCH₂CO₂Et →[DBU, THF/IPA] **3** (3,4-diethyl-2-ethoxycarbonylpyrrole) →[NaOH / (CH₂OH)₂, 190°C / 30 min] **4** (3,4-diethylpyrrole)

D. **4** + CH₂O$_{(aq)}$ →[1) C₆H₆/p-TsOH/reflux; 2) O₂] **5** (octaethylporphyrin)

Submitted by Jonathan L. Sessler,[1] Azadeh Mozaffari, and Martin R. Johnson.
Checked by Jürgen Fischer and Ekkehard Winterfeldt.

1. Procedure

A. 4-Nitro-3-hexanol (**1**).[2] To a 2-L, three-necked, round-bottomed flask equipped with a mechanical stirrer, thermometer, dropping funnel, and drying tube are added propionaldehyde (174 g, 3 mol) and isopropyl alcohol (IPA) (450 mL) (Note 1). The solution is stirred while finely ground potassium fluoride (25 g, 0.15 mol) is added to the flask. 1-Nitropropane (267.3 g, 3 mol) (Note 1) is then added dropwise with stirring, and the temperature is kept below 40°C with the aid of an ice bath (Note 2). The ice bath is removed about 30 min after the addition of 1-nitropropane is complete. The flask contents are stirred for an additional 18 hr. The catalyst is then removed by filtration and the filtrate is concentrated under reduced pressure. The residue is poured into water (500 mL) and the oil is extracted with ether (3 x 300 mL). The ethereal layer is dried over anhydrous sodium sulfate (Na_2SO_4), and the solvent is removed under reduced pressure. The remaining liquid is distilled under reduced pressure and the fraction boiling at 88-90°C/2 mm is collected in a tared, 1-L round-bottomed flask, yielding 3-nitro-4-hexanol (330 g, 2.24 mol, 65%) (Note 3). The flask containing the product is used directly in the next step.

B. 4-Acetoxy-3-nitrohexane (**2**).[3] To the above flask, containing 3-nitro-4-hexanol (330 g, 2.24 mol), is added a magnetic stirring egg and 1 mL of concd sulfuric acid. The contents of the flask are stirred in an ice bath and acetic anhydride (240 g, 2.35 mol) is added in portions, keeping the temperature of the reactants below 60°C. After the addition of the acetic anhydride is complete, the contents of the flask are stirred for 1 hr. The flask is then equipped for vacuum distillation. The lower boiling components (Ac_2O and AcOH) are removed at 25 mm by gently heating the stirred contents of the flask (≤ 100°C bath temperature). After these reagents have been removed, the system is cooled, attached to a vacuum pump, and carefully heated. The fraction boiling at 105-107°C/10 mm is collected, affording 4-acetoxy-3-nitrohexane (379 g, 2.0 mol, 90%) (Note 4).

C. *Ethyl 3,4-diethylpyrrole-2-carboxylate* (**3**)[4] *and 3,4-Diethylpyrrole* (**4**). A 1-L, three-necked, round-bottomed flask is equipped with a magnetic stirring bar, dropping funnel, thermometer, and drying tube, and charged with 3-acetoxy-2-nitrohexane (103 g, 0.54 mol), ethyl isocyanoacetate (50.7 g, 0.45 mol, Note 5), anhydrous tetrahydrofuran (320 mL), and anhydrous isopropyl alcohol (IPA) (130 mL) (Note 1). 1,8-Diazabicyclo[5.4.0]undec-7-ene (DBU, 152 g, 1 mol) (Notes 6 & 7) is then added, taking care to maintain the temperature at 20°C to 30°C at all times with the aid of an ice bath (Note 8). When addition of DBU is complete, the orange solution is stirred for 4 hr at room temperature. The solvent is completely removed under reduced pressure (50°C bath temp, 20-40 mm) and the residue is poured into a 1-L beaker and diluted with warm water (300 mL). To this biphasic mixture is added diethyl ether (300 mL). The contents of the beaker are poured into a separatory funnel. The aqueous layer is drawn off and extracted with an additional two portions of ether (300 mL). The ether layers are combined and washed with aqueous 10% hydrochloric acid (2 x 300 mL) and dried over magnesium sulfate ($MgSO_4$). The ether is removed under reduced pressure in a 1-L round-bottomed flask, leaving approximately 95 g of crude ethyl 3,4-diethylpyrrole-2-carboxylate (**3**) (Note 7). This material is not isolated, but is decarboxylated directly as follows: To the crude product **3** (95 g) is added sodium hydroxide (30 g, 0.75 mol) and ethylene glycol (300 mL). The contents are held at reflux under nitrogen for 1 hr, cooled, transferred to a 2-L separatory funnel, and diluted with water (500 mL) and hexane (600 mL). The layers are separated, and the aqueous layer is extracted further with hexane (3 x 300 mL). The hexane layers are combined, dried over $MgSO_4$, and concentrated under reduced pressure. The residue is distilled under reduced pressure, and the fraction boiling at

100°C/25 mm is collected, yielding 3,4-diethylpyrrole (21.14-22.00 g, 0.17-0.177 mol, 38.1-40%) (Note 9).

D. *2,3,7,8,12,13,17,18-Octaethylporphyrin* (**5**). A 500-mL, round-bottomed flask is wrapped with aluminum foil and equipped with a reflux condenser with a Dean-Stark trap, mechanical stirrer, and nitrogen inlet. The flask is charged with 3,4-diethylpyrrole (1 g, 8.1 mmol), benzene (300 mL) (Note 10), a 37% solution of aqueous formaldehyde (0.73 mL, 8.9 mmol), and *p*-toluenesulfonic acid (0.03 g, 1.7 mmol). The mixture is stirred and heated at reflux under nitrogen using an oil bath, and the water is removed by means of the Dean-Stark trap. After 8 hr, the solution is cooled, and the Dean-Stark trap and condenser are replaced with a fritted glass aerator/bubbler. Oxygen is bubbled through the brown mixture while it is stirred for 12-24 hr. Benzene is removed from the flask by distillation under reduced pressure, and the residue is dissolved in chloroform (20 mL) (Note 11). The solution is washed with 1 N sodium hydroxide (40 mL) and water (2 x 20 mL). The chloroform solution is concentrated to 5 mL in a 100-mL, round-bottomed flask, carefully layered over with methanol (≈70 mL), and allowed to stand for 48 hr. The resulting solid is collected by filtration and dried under reduced pressure for 24 hr. The crude material is recrystallized twice from chloroform-hexanes [effected by dissolving in chloroform (≈10 mL), layering over with hexanes (≈70 mL), and allowing to stand overnight]. The final precipitate is collected by filtration and dried under reduced pressure for 48 hr to yield analytically pure 2,3,7,8,12,13,17,18-octaethylporphyrin (720 mg, 1.34 mmol, 66.4%) as a purple, amorphous powder (Note 12).

2. Notes

1. Propionaldehyde and 1-nitropropane were obtained from Aldrich Chemical Company, Inc., and used as received. Isopropyl alcohol and tetrahydrofuran were obtained from J.T. Baker and used as received.

2. It is necessary to cool the reaction vessel to prevent the volatile propionaldehyde from evaporating.

3. The literature boiling point is reported[2] as 89°C (2 mm).

4. The spectral and analytical properties are as follows: ^{1}H NMR (CDCl$_3$, 300 MHz) δ: 0.99 (m, 6 H, CH$_3$), 1.62 and 1.80 (2 x m, 2 H, O$_2$NCHC$\underline{H}_2$CH$_3$), 1.99 and 2.12 (2 x m, 2 H, CH$_3$CO$_2$CHC$\underline{H}_2$CH$_3$), 2.06 (m, 3 H, CH$_3$CO$_2$), 4.56 (m, 1 H, C$\underline{H}$NO$_2$), 5.16 and 5.24 (2 x m, 1 H, CH$_3$CO$_2$C$\underline{H}$); C.I. MS, (M+1)$^+$ 190 (calcd for C$_8$H$_{15}$NO$_4$·H: 190). Anal. Calcd for C$_8$H$_{15}$NO$_4$: C, 50.78; H, 7.99; N, 7.40. Found: C, 50.98; H, 8.14; N, 7.01.

5. A disadvantage of the present procedure is that it requires the use of the relatively foul-smelling substance, ethyl isocyanoacetate. Although this material is commercially available (from, e.g., Aldrich Chemical Company, Inc.), it is moderately expensive. The authors have found that the existing preparative procedure (Hartman, G. D.; Weinstock, L. M. *Org. Synth., Coll. Vol VI* **1988**, 620) can be improved by the use of trichloromethyl chloroformate (Kurita, K.; Iwakura, Y. *Org. Synth., Coll. Vol. VI* **1988**, 715) rather than phosphoryl chloride. This substitution simplifies purification of the isocyanoacetate by eliminating the aqueous portion of the workup.

6. DBU was obtained from Aldrich Chemical Company, Inc. and used as received.

7. Two equivalents of DBU are used here. One equivalent of DBU eliminates acetate from one of the reactants to form 3-nitro-3-hexene in situ, which goes on to form the pyrrole. The intermediate ethyl 3,4-diethylpyrrole-2-carboxylate can also be

prepared directly from ethyl isocyanoacetate and 3-nitro-3-hexene in good yield (86%) under conditions similar to those outlined here.[5] Although this alternative requires a further manipulative step, it requires only half as much DBU.

8. It is important not to allow the temperature to drop below 20°C because the reaction slows down considerably. Unreacted DBU then builds up. As a result, when the temperature does climb, it does so rapidly (often to as high as 65°C). This results in a significantly lower yield.

9. The spectral and physical properties are as follows: ^{1}H NMR (CDCl$_3$, 300 MHz) δ: 1.16 (t, 6 H, CH$_2$C$\underline{H}_3$), 2.47 (q, 4 H, C$\underline{H}_2$CH$_3$), 6.42 (d, 2 H, pyrrole CH), 7.65 (s, 1 H, pyrrole NH); MS m/e (relative intensity) 123 (46), 108 (100), 93 (37); bp 100°C/25 mm; 69°C/7 mm (lit.[6] bp, 83°C/10 mm).

10. Benzene is a known carcinogen. Follow manufacturer's recommended procedures for handling, storage, and disposal.

11. Chloroform is a suspected carcinogen. Follow manufacturer's recommended procedures for handling, storage, and disposal.

12. The spectral and analytical properties are as follows: ^{1}H NMR (CDCl$_3$, 300 MHz) δ: -3.72 (s, 2 H, NH), 1.95 (t, 24 H, CH$_2$C$\underline{H}_3$), 4.12 (q, 16 H, C$\underline{H}_2$CH$_3$), 10.12 (s, 4 H, meso CH); HRMS, M$^+$ 534.37351 (calcd for C$_{36}$H$_{46}$N$_4$: 534.37225). Anal. Calcd for C$_{36}$H$_{46}$N$_4$: C, 80.85; H, 8.67; N, 10.48. Found: C, 80.89; H, 8.56; N, 10.37; UV-vis (CHCl$_3$-MeOH 95:5 vv.) λ_{max} (log ε): 398 (5.20), 498 (4.10), 533 (4.00), 565 (3.79), 618 (3.68) nm.

Waste Disposal Information

All toxic materials were disposed of in accordance with "Prudent Practices for Disposal of Chemicals from Laboratories"; National Academy Press; Washington, DC, 1983.

3. Discussion

Octaethylporphyrin (OEP) and tetraphenylporphyrin (TPP) remain among the most widely used of an increasingly diverse set of available synthetic porphyrins. The inherently high symmetry and relatively good solubility properties of these systems often combine to make them the models of choice for a wide range of biological modeling and inorganic chemical applications.[7] Recently, an optimized synthesis of TPP and related tetraarylporphyrins has been developed by Lindsey and co-workers.[8] At present, however, the synthesis of OEP (5) remains problematic: Although numerous strategies have been reported,[5,9-14,15] no convenient, high-yield procedure currently exists.

Traditionally, octaethylporphyrin has been prepared by the self-condensation of 2-N,N'-diethylaminomethyl-3,4-diethylpyrrole,[9,10] ethyl 5-N,N'-diethylaminomethyl-3,4-diethylpyrrole-2-carboxylate,[11,12] or 3,4-diethyl-5-hydroxymethylpyrrole-2-carboxylic acid under oxidative conditions.[13] It has also been prepared on a small scale directly from 3,4-diethylpyrrole in 65% yield by condensation with aqueous formaldehyde under acid-catalyzed conditions,[14] using conditions similar to those which have proved useful for preparing the corresponding octamethylporphyrin analogue.[16] All of these syntheses derive from the same, initial pyrrole precursor, namely, ethyl 3,4-diethyl-5-methylpyrrole-2-carboxylate, prepared from the classic, reverse-sense Knorr reaction of ethyl propionylacetate with 2,4-pentanedione, and they require several steps before the ultimate porphyrin-forming condensation. Octaethylporphyrin has also been prepared recently by the reduction of 2,8,12,18-tetraacetyl-3,7,13,17-tetraethylporphyrin by diborane,[14] and by the condensation of 3,4-diethylpyrrole-N-carboxylic acid with formaldehyde in refluxing acetic acid/pyridine.[15] Neither of these procedures, however, truly overcomes the problem associated with preparing the initial pyrrole.

The synthesis reported here circumvents many of the problems associated with existing preparative methods. Specifically, it makes use of a new procedure of Barton and Zard[4] in the key pyrrole-forming step. This method, which gives an α-unsubstituted pyrrole ester (e.g., **3**) directly in good yield, provides a substantial saving in labor when compared to the Knorr approach, and it is very flexible with regard to the kinds of β-substitution allowed. Since the remaining α-ester group can be conveniently removed by saponification and subsequent decarboxylation (often, as is the case here, without isolation of the initial pyrrole product), this method provides a quick and easy means of preparing 3,4-dialkylated pyrroles. Simple acid-catalyzed condensation of the resulting 3,4-dialkylpyrroles with formaldehyde and subsequent oxidation is then all that is required to complete the synthesis of an octaalkylporphyrin.[17,18] We have found that these latter transformations may be readily effected using aqueous formaldehyde under acid-catalyzed dehydrating conditions, followed by simple air-induced oxidation. In the specific case of octaethylporphyrin, when the reaction is run on a 1-g scale, a 75% yield of analytically pure product is obtained following workup and purification (which involves only simple recrystallizations and no chromatographic separations). This procedure can be conveniently scaled up by a factor of ten. Under these conditions, it still gives a good yield (55%) of pure product. It does, however, require relatively large amounts of benzene (3 L for a reaction carried out with 10 g of 3,4-diethylpyrrole), which could present a health hazard. However, if due caution is exercised with regard to this point, the present method provides an easy way to prepare large quantities of octaethylporphyrin. As such it represents a considerable advance over earlier methods in terms of both ease and convenience.

1. Department of Chemistry, University of Texas at Austin, Austin, TX 78712.
2. Procedure of Kambe and Yasuda: Kambe, S.; Yasuda, H. *Bull. Chem. Soc. Jpn.* **1968**, *41*, 1444.
3. Procedure of Tindall: Tindall, J. B. *Ind. Eng. Chem.* **1941**, *33*, 65.
4. Procedure of Barton and Zard: Barton, D. H. R.; Zard, S. Z. *J. Chem. Soc., Chem. Commun.* **1985**, 1098.
5. Ono, N.; Maruyama, K. *Chem. Lett.* **1988**, 1511.
6. Fischer, H.; Guggemos, H.; Schäfer, A. *Ann. Chem.* **1939**, *540*, 30.
7. "The Porphyrins"; Dolphin, D., Ed.; Academic Press: New York, 1978-1979; Vols. I-VII.
8. (a) Lindsey, J. S.; Hsu, H. C.; Schreiman, I. C. *Tetrahedron Lett.* **1986**, *27*, 4969; (b) Lindsey, J. B.; Schreiman, I. C.; Hsu, H. C.; Kearney, P. C.; Marguerettaz, A. M. *J. Org. Chem.* **1987**, *52*, 827; (c) Wagner, R. W.; Lawrence, D. S.; Lindsey, J. S. *Tetrahedron Lett.* **1987**, *28*, 3069.
9. (a) Eisner, U.; Lichtarowicz, A.; Linstead, R. P. *J. Chem. Soc.* **1957**, 733; (b) Eisner, U.; Linstead, R. P.; Parkes, E. A.; Stephen, E. *J. Chem. Soc.* **1956**, 1655.
10. Whitlock, H. W.; Hanauer, R. *J. Org. Chem.* **1968**, *33*, 2169.
11. Paine, J. B., III; Kirshner, W. B.; Moskowitz, D. W.; Dolphin, D. *J. Org. Chem.* **1976**, *41*, 3857.
12. Wang, C.-B.; Chang, C. K. *Synthesis*, **1979**, 548.
13. Inhoffen, H. H.; Fuhrhop, J. H.; Voight, H.; Brockmann, H., Jr. *Ann. Chem.* **1966**, *695*, 133.
14. (a) Cheng, D. O.; LeGoff, E. *Tetrahedron Lett.* **1977**, 1469; b) LeGoff, E.; Cheng, D. O., In "Porphyrin Chemistry Advances"; F. R. Longo, Ed.; Ann Arbor Science Publishers, 1979; pp. 153-156.

15. Callot, H. J.; Louati, A.; Gross, M. *Bull. Soc. Chim. Fr.* **1983**, 317; Callot, H. J.; Louati, A.; Gross, M. *Angew. Chem., Inter. Ed. Engl.* **1982**, *21*, 285.
16. Treibs, A.; Häberle, N. *Justus Liebigs Ann. Chem.* **1968**, *718*, 183.
17. Alternatively, the ethyl 3,4-diethylpyrrole-2-carboxylate may be carried on directly to give octaethylporphyrin,[5] although the yields reported (ca. 40%) are not quite as good as those obtained by the present procedure. Similarly, this substance or the 3,4-diethylpyrrole produced by the present procedure could conceivably serve as the basis for an an improved synthesis via a Mannich base-type approach such as that outlined in refs. 9-12.
18. The procedure reported here appears to be quite general. We have, for example, used it to prepare a β-substituted tetrakis-fused cyclohexylporphyrin (1,2,3,4,8,9,10,11,15,16,17,18,22, 23,24,25-hexadecahydro-29H,31H-tetra-benzo[BGLQ]porphine) in 51% overall yield starting from 1-nitrocyclohexene.

Appendix

Chemical Abstracts Nomenclature (Collective Index Number); (Registry Number)

3,4-Diethylpyrrole: Pyrrole, 3,4-diethyl- (8,9); (16200-52-5)

2,3,7,8,12,13,17,18-Octaethylporphyrin: Porphine, 2,3,7,8,12,13,17,18-octaethyl- (8); 21H,23H-Porphine, 2,3,7,8,12,13,17,18-octaethyl- (9); (2683-82-1)

4-Nitro-3-hexanol: 3-Hexanol, 3-nitro- (8,9); (5342-71-2)

Propionaldehyde (8); Propanal (9); (123-38-6)

1-Nitropropane: Propane, 1-nitro- (8,9); (108-03-2)

4-Acetoxy-3-nitrohexane: 3-Hexanol, 4-nitro-, acetate (9); (3750-83-2)

Acetic anhydride (8); Acetic acid anhydride (9); (108-24-7)

Ethyl 3,4-diethylpyrrole-2-carboxylate: 1H-Pyrrole-2-carboxylic acid, 3,4-diethyl-, ethyl ester (11); (97336-41-9)

Ethyl isocyanoacetate: Acetic acid, isocyano-, ethyl ester (8,9); (2999-46-4)

1,8-Diazabicyclo[5.4.0]undec-7-ene: Pyrimido[1,2-a]azepine, 2,3,4,6,7,8,9,10-octahydro- (8,9); (6674-22-2)

Formaldehyde (8,9); (50-00-0)

p-Toluenesulfonic acid (8); Benzenesulfonic acid, 4-methyl- (9); (6192-52-5)

Trichloromethyl chloroformate: Formic acid, chloro-, trichloromethyl ester (8); Carbonochloridic acid, trichloromethyl ester (9); (503-38-8)

PREPARATION AND DIELS-ALDER REACTION OF A REACTIVE, ELECTRON-DEFICIENT HETEROCYCLIC AZADIENE: DIMETHYL 1,2,4,5-TETRAZINE-3,6-DICARBOXYLATE. 1,2-DIAZINE AND PYRROLE

INTRODUCTION

(1,2,4,5-Tetrazine-3,6-dicarboxylic acid, dimethyl ester)

A. $2\ N_2CHCO_2C_2H_5 \xrightarrow[H_2O]{NaOH}$ [tetrazine-3,6-di(CO$_2$Na), dihydro] **1**

B. $\mathbf{1} \xrightarrow[H_2O]{HCl}$ [tetrazine-3,6-di(CO$_2$H), dihydro] **2**

C. $\mathbf{2} \xrightarrow[CH_3OH]{SOCl_2}$ [tetrazine-3,6-di(CO$_2$CH$_3$), dihydro] **3**

D. $\mathbf{3} \xrightarrow[CH_2Cl_2]{\text{nitrous gases}}$ [tetrazine-3,6-di(CO$_2$CH$_3$)] **4**

E. $\mathbf{4}\ +$ PhC(OSiMe$_3$)=CH$_2$ $\xrightarrow[25°C]{\text{dioxane}}$ [4-Ph-pyridazine-3,6-di(CO$_2$CH$_3$)] **5**

F. $\mathbf{5} \xrightarrow[CH_3COOH]{Zn}$ [3-Ph-pyrrole-2,5-di(CO$_2$CH$_3$)]

Submitted by Dale L. Boger, James S. Panek, and Mona Patel.[1]
Checked by Richard Hutchings and Albert I. Meyers.

1. Procedure[2]

A. Disodium dihydro-1,2,4,5-tetrazine-3,6-dicarboxylate. A 2-L, three-necked, round-bottomed flask is equipped with an overhead stirrer, thermometer, and a 500-mL addition funnel. Sodium hydroxide (320 g, 8 mol) and 500 mL of water are added. Ethyl diazoacetate (200 g, 1.75 mol, Note 1) is placed in the addition funnel and added dropwise to the stirred sodium hydroxide solution so as to maintain the temperature of the reaction mixture between 60°C and 80°C (approximately 1.5 hr, Note 2). After the reaction slurry is cooled to room temperature, it is poured into 2 L of 95% ethanol, mixed well, and the liquid is decanted. This washing procedure is repeated five times using 1.5 L of 95% ethanol each time. The precipitate is collected by filtration using a Büchner funnel, the collected solid is washed with 1 L of absolute ethanol and 1 L of ether, and dried (12 hr) in the air to afford 160-184 g (85-97%) of disodium dihydro-1,2,4,5-tetrazine-3,6-dicarboxylate as yellow brown solid.

B. Dihydro-1,2,4,5-tetrazine-3,6-dicarboxylic acid. A 2-L, three-necked, round-bottomed flask is equipped with an overhead stirrer and a 100-mL addition funnel. Disodium dihydro-1,2,4,5-tetrazine-3,6-dicarboxylate (90.37 g 0.42 mol) in 100 mL of water and 100 g of crushed ice are added. The resulting slurry is cooled with an ice/sodium chloride bath and a solution of concentrated hydrochloric acid (84 mL of 36-38%) is added dropwise with stirring over 45 min. The reaction mixture is washed five times with 200 mL of dry ether and the ether layer is decanted. The product is immediately collected by filtration using a Büchner funnel. Drying the collected solid at room temperature under reduced pressure affords 51.6-53.06 g (72-74%) of dihydro-1,2,4,5-tetrazine-3,6-dicarboxylic acid as a yellow powder: mp 144-148°C (Note 3).

C. Dimethyl dihydro-1,2,4,5-tetrazine-3,6-dicarboxylate. Absolute methanol (700 mL, Note 4) is placed in a dry, 2-L, round-bottomed flask fitted with an overhead stirrer, thermometer, and a 100-mL addition funnel and is cooled to -30°C. Thionyl chloride (62.12 g, 0.522 mol, 38.1 mL, Note 5) is added carefully with stirring and the reaction mixture is allowed to stir at -30°C for 30 min. Dihydro-1,2,4,5-tetrazine-3,6-dicarboxylic acid (45.00 g, 0.261 mol) is added as a solid in portions over 30 min to the cooled, stirred thionyl chloride-methanol solution (Note 6). The reaction mixture is allowed to warm to room temperature (1 hr) and is subsequently warmed to 35-40°C (internal temperature) for 2 hr. The reaction mixture is cooled to -30°C and the precipitate is collected by filtration using a Büchner funnel. The collected solid is washed with ether (115 mL, Note 7) and the filtrate is concentrated under reduced pressure to give an orange-brown oil (ca. 15 g). The collected solid is triturated with methylene chloride (2.0 L) and the insoluble inorganic salts are removed by filtration using a Büchner funnel. The orange-brown oil (ca. 15 g) is taken up in water (150 mL) and extracted with methylene chloride (6 x 270 ml). The combined methylene chloride extracts and methylene chloride triturate are dried over anhydrous sodium sulfate and concentrated under reduced pressure to afford 23-31 g (44-51%) of pure dimethyl dihydro-1,2,4,5-tetrazine-3,6-dicarboxylate as an orange-yellow powder: mp 171-172°C (Note 8).

D. *Dimethyl 1,2,4,5-tetrazine-3,6-dicarboxylate.* Dimethyl dihydro-1,2,4,5-tetrazine-3,6-dicarboxylate (20 g, 0.1 mol) is slurried in 800 mL of methylene chloride (Note 9) in a 2-L, round-bottomed flask fitted with a magnetic stirring bar and the mixture is cooled with an ice/water bath. A stream of nitrous gases (Note 10) is bubbled into the reaction mixture with stirring for 15 min. The color of the reaction mixture changes from orange to bright red during the addition. Stirring is continued for 1.5 hr as the reaction mixture is allowed to warm to room temperature. The solvent and excess nitrous gases are removed under reduced pressure to afford dimethyl 1,2,4,5-tetrazine-3,6-dicarboxylate (19.7 g, 99%) as a bright red, crystalline solid: mp 173-175°C (Note 11).

E. Dimethyl 4-phenyl-1,2-diazine-3,6-dicarboxylate. A 50-mL, round-bottomed flask equipped with a magnetic stirring bar is charged with dimethyl 1,2,4,5-tetrazine-3,6-dicarboxylate (1.0 g, 0.005 mol) and 1,4-dioxane (20 mL, Note 12). 1-Phenyl-1-(trimethylsiloxy)ethylene (1.07 g, 0.0056 mol, 1.14 mL, Note 13) is added and the reaction mixture is stirred under nitrogen at room temperature for 8 hr. The solvent is removed under reduced pressure to give a viscous oil that is triturated with anhydrous ether (2-3 mL). The solid product is collected by vacuum filtration and recrystallized from ethyl acetate/hexane to give 1.23-1.30 g (90-96%) of dimethyl 4-phenyl-1,2-diazine-3,6-dicarboxylate as a light yellow solid, mp 95.5-96°C (Note 14).

F. Dimethyl 3-phenylpyrrole-2,5-dicarboxylate. A 250-mL, round-bottomed flask equipped with a magnetic stirring bar is charged with dimethyl 4-phenyl-1,2-diazine-3,6-dicarboxylate (1.36 g, 0.005 mol) and glacial acetic acid (55 mL, Note 15). Zinc dust (3.25 g, 0.05 mol, Note 16) is added and the reaction mixture is stirred at room temperature for 6 hr. A second portion of zinc dust (3.25 g, 0.05 mol) is added and the reaction mixture is stirred for an additional 18 hr. The zinc dust is removed by filtration through a pad of Celite and the residue is washed with ether (100 mL). The filtrate and washes are combined, made basic (pH 10) with the addition of saturated sodium bicarbonate, and extracted with ether (2 x 100 mL). The combined ether extracts are dried over anhydrous magnesium sulfate and concentrated under reduced pressure. Purification of the product is effected by flash chromatography on a 4.5 x 9-cm column of silica gel (Aldrich 951, CH_2Cl_2 eluant), collecting 20-mL fractions. The fractions are analyzed by thin-layer chromatography (Kieselgel 60, CH_2Cl_2 eluant) and those containing product are combined and concentrated in vacuo to give 0.676 g (52%) of dimethyl 3-phenylpyrrole-2,5-dicarboxylate as a white solid: mp 122-123°C (ethyl acetate-hexane, Note 17).

2. Notes

1. The submitters employed, without purification, ethyl diazoacetate obtained from Aldrich Chemical Company, Inc.

2. A time lag (10-15 min) is observed before the exothermic reaction begins. Addition of ethyl diazoacetate is then maintained at such a rate that the reaction temperature does not rise above 80°C. The checkers had to heat the mixture to 60°C.

3. Drying of the free acid should be rapid with a large surface area since traces of hydrochloric acid promote hydrolysis of the product to hydrazine salts. Slight warming (≤ 60°C) during drying accelerates the drying process. The IR spectrum is as follows: IR (KBr) γ_{max} cm^{-1}: 3700-3100, 3320, 3000-1850, 1710. 1630. The checkers found that this step does not work as well on a smaller scale (0.14 mol).

4. Methanol is distilled from magnesium turnings immediately before use.

5. The submitters employed, without purification, thionyl chloride obtained from Fisher Scientific Company. The procedure should be performed in a well-ventilated hood since thionyl chloride is a lachrymator. The yield of dimethyl ester was found to be lower in instances when the thionyl chloride-methanol solution was not allowed to stir (30 min, -30°C) prior to the addition of dihydro-1,2,4,5-tetrazine-3,6-dicarboxylic acid.

6. The temperature is maintained at -30°C during the additions.

7. The submitters employed ether distilled from sodium benzophenone ketyl.

8. The spectral properties of the product are as follows: ^{1}H NMR (CDCl$_3$) δ: 3.92 (s, 6 H, CO$_2$CH$_3$), 7.42 (br s, 2 H, NH); IR (KBr) v_{max} cm^{-1}: 3160, 3050, 1740, 1720.

9. The submitters employed methylene chloride from Fisher Scientific Company, which was distilled before use.

10. Nitrous gases are generated in a separate vessel by the disproportionation of nitrous acid (HONO): 200 mL of 6 N NaNO$_2$ (1.2 mol) is added dropwise to 125 mL of concentrated hydrochloric acid (1.5 mol) in a 500-mL, three-necked, round-bottomed flask fitted with a nitrogen inlet, a 500-mL addition funnel, and an outlet tube leading to the reaction flask. The brown gases evolved are bubbled directly into the reaction mixture through a 5-mm (inside diameter) glass tube (smaller inlet tubes occasionally became plugged) using a nitrogen stream. *Caution: all operations involving nitrous gases should be conducted in a well-ventilated hood because of the toxicity of these gases.*

11. The checkers observed some starting material in the product which depressed the mp. It could be removed by crystallization from ethyl acetate to give pure **3**, mp 176-177°C, but with significant loss of product. The spectral properties of the product are as follows: ^{1}H NMR (CDCl$_3$) δ: 4.22 (s, 6 H, CO$_2$CH$_3$); IR (KBr) ν$_{max}$ cm^{-1}: 2970, 1752, 1445, 1385, 1219, 1175, 1082, 960, 912; UV (dioxane) λ$_{max}$ (log ε) 520 nm (2.754).

12. The submitters employed 1,4-dioxane obtained from Fisher Scientific Company and distilled before use.

13. The submitters employed, without purification, 1-phenyl-1-(trimethylsiloxy)ethylene obtained from Aldrich Chemical Company, Inc.

14. The elemental analysis and the spectral analysis of the product are as follows: Anal. Calcd for C$_{14}$H$_{12}$N$_2$O$_4$: C, 61.76; H, 4.44; N, 10.29. Found: C, 62.01; H, 4.50; N, 10.19; ^{1}H NMR (CDCl$_3$) δ: 3.89 (s, 3 H, CO$_2$CH$_3$), 4.12 (s, 3 H, CO$_2$CH$_3$), 7.40-7.60 (m, 5 H, Ph), 8.27 (s, 1 H, C5-H); IR (KBr) ν$_{max}$ cm^{-1}: 2955, 1742, 1584, 1447, 1399, 1287, 1244, 1142, 766; EI-MS (70 eV): m/e (relative intensity) 272 (M+, 9), 242 (7), 241 (6), 214 (34), 182 (10), 155 (base), lit[2b] mp 94-95.5°C.

15. The submitters employed, without purification, glacial acetic acid obtained from Fisher Scientific Company.

16. Zinc dust obtained from Fisher Scientific Company was activated prior to use following an established procedure.[3]

17. The elemental analysis and the spectral analysis of the product are as follows: Anal. Calcd for $C_{14}H_{13}NO_4$: C, 64.86; H, 5.05; N, 5.40. Found: C, 65.10; H, 4.99; N, 5.48; ^{1}H NMR (CDCl$_3$) δ: 3.82 (s, 3 H, OCH$_3$), 3.91 (s, 3 H, OCH$_3$), 6.94 (d, 1 H, J = 3, C4-H), 7.30-7.60 (m, 5 H, Ph), 9.80 (br, s, 1 H, NH); IR (KBr) v_{max} cm^{-1}: 3314, 2958, 1726, 1564, 1464, 1436, 1270, 1096, 1008, 940, 846. 762, 696. The additional formation of methyl 3-phenyl-5-carboxamidopyrrole-2-carboxylate in 32% yield is observed. The spectral properties of the product are as follows: ^{1}H NMR (CDCl$_3$) δ: 1.75 (br s, 2 H), 3.92 (s, 3 H), 7.14 (s, 1 H), 7.40-7.60 (m, 5 H); IR (KBr) v_{max} cm^{-1}: 3442, 3346, 2950, 1708, 1642, 1580, 1528, 1476, 1366, 1280, 1132, 1022, 936, 808, 728, 618; CI-MS (70 eV): m/e (relative intensity) 245 (M$^+$ + H, base).

Waste Disposal Information

All toxic materials were disposed of in accordance with "Prudent Practices for Disposal of Chemicals from Laboratories"; National Academy Press; Washington, DC, 1983.

3. Discussion

The procedure describes the preparation and use of a reactive, electron-deficient heterocyclic azadiene suitable for Diels-Alder reactions with electron-rich, unactivated, and electron-deficient dienophiles.[4] Dimethyl 1,2,4,5-tetrazine-3,6-dicarboxylate, because of its electron-deficient character, is ideally suited for use in inverse electron demand (LUMO$_{diene}$-controlled)[5] Diels-Alder reactions. Table I and Table II detail representative examples of the reaction of dimethyl 1,2,4,5-tetrazine-3,6-

dicarboxylate with electron-rich carbon dienophiles[6d] and heterodienophiles,[4a,b,c] respectively. Complete surveys of the reported Diels-Alder reactions of dimethyl 1,2,4,5-tetrazine-3,6-dicarboxylate have been compiled.[4] Reductive ring contraction of the substituted dimethyl 1,2-diazine-3,6-dicarboxylate [4 + 2] cycloadducts effected by zinc in acetic acid provides the corresponding substituted dimethyl pyrrole-2,5-dicarboxylates.[6d,7] Table III details representative examples of this general reductive ring contraction reaction.[7,8]

This approach to 1,2-diazine and pyrrole introduction based on the inverse electron demand Diels-Alder reaction of dimethyl 1,2,4,5-tetrazine-3,6-dicarboxylate complements the [4 + 2] cycloaddition reactions of a range of electron-deficient heterocyclic azadienes which permits the divergent preparation of a range of heterocyclic agents employing a common dienophile precursor, Scheme I.

1. Department of Chemistry, Purdue University, West Lafayette, IN 47907.
2. The procedure described for the preparation of dimethyl 1,2,4,5-tetrazine-3,6-dicarboxylate is an adaptation of two detailed preparations: (a) Spencer, G. H., Jr.; Cross, P. C.; Wiberg, K. B. *J. Chem. Phys.* **1961**, *35*, 1939; (b) Sauer, J.; Mielert, A.; Lang, D.; Peter, D. *Chem. Ber.* **1965**, *98*, 1435. (c) The oxidation of dimethyl dihydro-1,2,4,5-tetrazine-3,6-dicarboxylate to dimethyl 1,2,4,5-tetrazine-3,6-dicarboxylate using nitrous gases is a modification of a previously described procedure,[2b] see: Boger, D. L.; Coleman, R. S.; Panek, J. S.; Huber, F. X.; Sauer, J. *J. Org. Chem.* **1985**, *50*, 5377. For the original description of the base-promoted dimerization of ethyl diazoacetate, see: Curtius, Th.; Darapsky, A.; Müller, E. *Chem. Ber.* **1906**, *39*, 3410; **1907**, *40*, 84; **1908**, *41*, 3161; Curtius, Th.; Lang, J. *J. Prakt. Chem.* **1888**, *38*, 531; Hantzsch, A.; Lehmann, M. *Chem. Ber.* **1900**, *33*, 3668; Hantzsch, A.; Silberrad, O. *Chem. Ber.* **1900**, *33*, 58.

3. Fieser, L. F.; Fieser, M. "Reagents for Organic Synthesis"; Wiley: New York, 1967; Vol. 1, p. 1276.
4. (a) Neunhoeffer, H.; Wiley, P. F. "The Chemistry of Heterocyclic Compounds" In "Chemistry of 1,2,3-Triazines and 1,2,4-Triazines, Tetrazines, and Pentazines"; Wiley: New York; 1978, Vol. 33; (b) Boger, D. L. *Tetrahedron* **1983**, *39*, 2869; (c) Boger, D. L. *Chem. Rev.* **1986**, *86*, 781; (d) Boger, D. L.; Weinreb, S. M. "Hetero Diels-Alder Methodology in Organic Synthesis"; Academic Press, FL; 1987; (e) Weinreb, S. M.; Staib, R. R. *Tetrahedron* **1982**, *38*, 3087.
5. Houk, K. N. *J. Am. Chem. Soc.* **1973**, *95*, 4092.
6. (a) Boger, D. L.; Panek, J. S. *J. Org. Chem.* **1983**, *48*, 621; (b) Boger, D. L.; Panek, J. S. *Tetrahedron Lett.* **1983**, *24*, 4511; (c) Boger, D. L.; Panek, J. S. *J. Am. Chem. Soc.* **1985**, *107*, 5745; (d) Boger, D. L.; Coleman, R. S.; Panek, J. S.; Yohannes, D. *J. Org. Chem.* **1984**, *49*, 4405.
7. Bach, N. J; Kornfeld, E. C.; Jones, N. D.; Chaney, M. O.; Dorman, D. E.; Paschal, J. W.; Clemens, J. A.; Smalstig, E. B. *J. Med. Chem.* **1980**, *23*, 481.
8. For additional examples, see: Boger, D. L.; Patel, M. *J. Org. Chem.* **1988**, *53*, 1405.

Scheme 1

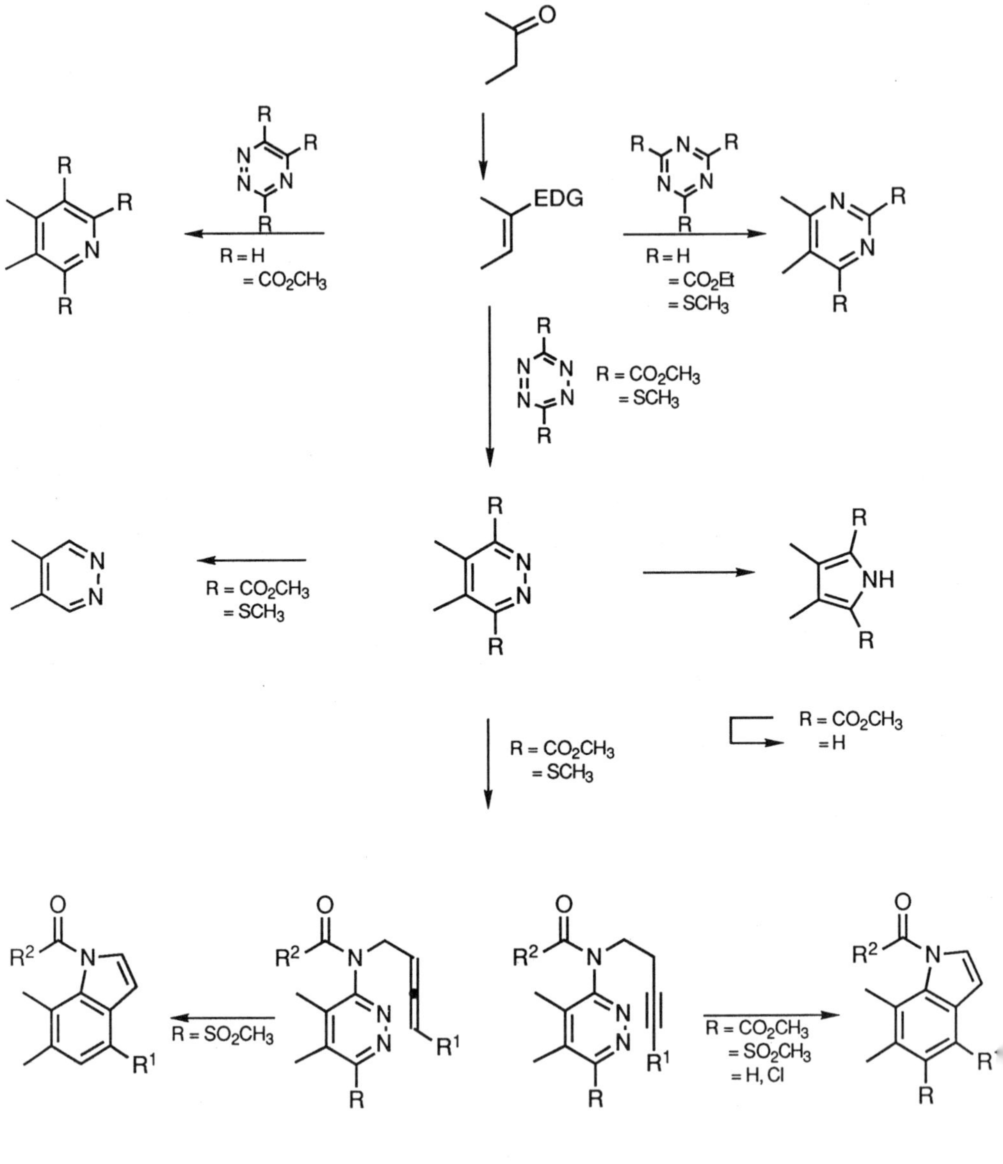

TABLE I

Diels-Alder Reactions of Dimethyl 1,2,4,5-Tetrazine-3,6-dicarboxylate: 1,2-Diazine Introduction[6d]

Entry	Dienophile	Conditions[a] Equiv. Temp °C (time hr)	1,2-Diazine		% Yield
1	CH₂=C(OSiEt₃)(CH₃)	1.5,25(12)	(dimethyl-substituted diazine)	1	87
2	Me—≡—Me	2-6,25(12)		1	trace
3	X = morpholine	2,25(48)	(ethyl/methyl diazine)	2	70
4	= pyrrolidine	2,25(48)			trace
5	1-pyrrolidinocyclohexene	1.5,25(12)	(tetrahydrophthalazine)	3	85
6	X = OSi(CH₃)₃	1,25(5)	Ph-substituted diazine	4	92
7	= morpholine	1.2,25(1.5)			87
8	= pyrrolidine	1.5,25(12)			trace
9	CH₃O–C(=CH₂)–OCH₃	1.5,25(0.5)	CH₃O-diazine	5	65
10	RO–C(=CH₂)–OCH₂Ph, R = Si(Me)₂-t-Bu	1.5,5-25(0.5)	PhCH₂O-diazine	6	33
11	PhCH₂O—≡—H	2-3,25(6)		6	82
12	CH₃O–C(=C(COCH₃))–OCH₃	2.5,101(3)	CH₃O/acetyl-diazine	7	71
13	CH₃O₂C-bridged bicyclic enamine with morpholine	1,25(5)	CH₃O₂C-bridged diazine	8	69

(a) All Diels-Alder reactions were carried out in dioxane.

TABLE II
Diels-Alder Reaction of Dimethyl 1,2,4,5-Tetrazine-3,6-dicarboxylate with C=N Heterodienophiles[6a-c]

Dienophile	Conditions Temp °C (time hr)[a]	Product	% Yield
(2-pyridyl C(=NH)X) X = SCH_3 = OEt = NH_2 = NEt_2	80(20) 80(8-12) 25(5) 25-50(25)	5-(2-pyridyl)-1,2,4-triazine-3,6-dicarboxylate	68 37 - -
(Ph C(=NH)X) X = SCH_3 = OEt = NH_2	80(24) 60(10) 25(5)	5-phenyl-1,2,4-triazine-3,6-dicarboxylate	65 27 -

(quinoline-2-C(=NH)X with R^1, R^2)
$R^1 = R^2 = H$
  X = SCH_3
  = OEt
$R^1 = OCH_3$ $R^2 = H$
  X = SCH_3
$R^1 = OCH_3$ $R^2 = NO_2$
  X = SCH_3
	80(4)
	80(20)
	
	80(4)
	
	80(20-24)
		(quinolinyl-substituted triazine product)	70
			33
			
			78
			
			82

(a) All Diels-Alder reactions were carried out in dioxane.

TABLE III
Reductive Ring Contraction of Substituted Dimethyl 1,2-Diazine-3,6-dicarboxylates: Pyrrole Introduction[6d]

Entry	1,2-Diazine	Conditions[a] Temp °C (time hr)	Pyrrole	%Yield
1	dimethyl 4,5-dimethyl-1,2-diazine-3,6-dicarboxylate	25(24)	dimethyl 3,4-dimethylpyrrole-2,5-dicarboxylate	63
2	dimethyl 4-ethyl-5-methyl-1,2-diazine-3,6-dicarboxylate	25(24)	dimethyl 3-ethyl-4-methylpyrrole-2,5-dicarboxylate	70
3	tetrahydrophthalazine dicarboxylate	25(22)	tetrahydroisoindole dicarboxylate	52
4	dimethyl 4-phenyl-1,2-diazine-3,6-dicarboxylate	25(9)	dimethyl 3-phenylpyrrole-2,5-dicarboxylate	65
5	dimethyl 4-methoxy-1,2-diazine-3,6-dicarboxylate	25(24)	dimethyl 3-methoxypyrrole-2,5-dicarboxylate	67
6	dimethyl 4-benzyloxy-1,2-diazine-3,6-dicarboxylate	25(24)	dimethyl 3-benzyloxypyrrole-2,5-dicarboxylate	62
7	dimethyl 4-methoxy-5-acetyl-1,2-diazine-3,6-dicarboxylate	25(24)	dimethyl 3-methoxy-4-acetylpyrrole-2,5-dicarboxylate	56
8	bridged CH$_3$O$_2$C-N diazine dicarboxylate	25(36)	bridged CH$_3$O$_2$C-N pyrrole dicarboxylate	48

(a) All zinc (9-20 molar equiv) reductions were carried out in acetic acid (0.09 M in substrate).

Appendix

Chemical Abstracts Nomenclature (Collective Index Number); (Registry Number)

Dimethyl 1,2,4,5-tetrazine-3,6-dicarboxylate: s-Tetrazine-3,6-dicarboxylic acid, dimethyl ester (8); 1,2,4,5-Tetrazine-3,6-dicarboxylic acid, dimethyl ester (9); (2166-14-5)

Disodium dihydro-1,2,4,5-tetrazine-3,6-dicarboxylate: 1,2,4,5-Tetrazine-3,6-dicarboxylic acid, 1,2-dihydro-, disodium salt (11); (96898-32-7)

Ethyl diazoacetate: Acetic acid, diazo-, ethyl ester (8,9); (623-73-4)

Dihydro-1,2,4,5-tetrazine-3,6-dicarboxylate: 1,2,4,5-Tetrazine-3,6-dicarboxylic acid, 1,2-dihydro- (9); (3787-09-5)

Dimethyl dihydro-1,2,4,5-tetrazine-3,6-dicarboxylate: 1,2,4,5-Tetrazine-3,6-dicarboxylic acid, 1,2-dihydro-, dimethyl ester (9); (3787-10-8)

Thionyl chloride (8,9); (7719-09-7)

Dimethyl 4-phenyl-1,2-diazine-3,6-dicarboxylate: 3,6-Pyridazinedicarboxylic acid, 4-phenyl-, dimethyl ester (9); (2166-27-0)

1-Phenyl-1-(trimethylsiloxy)ethylene: Silane, trimethyl[(1-phenylethenyl)oxy]- (9); (13735-81-4)

Dimethyl 3-phenylpyrrole-2,5-dicarboxylate: 1H-Pyrrole-2,5-dicarboxylic acid, 3-phenyl-, dimethyl ester (11); (92144-12-2)

SYNTHESIS OF FURANS VIA RHODIUM(II) ACETATE-CATALYZED REACTION OF ACETYLENES WITH α-DIAZOCARBONYLS: ETHYL 2-METHYL-5-PHENYL-3-FURANCARBOXYLATE

(3-Furancarboxylic acid, 2-methyl-5-phenyl-, ethyl ester)

A. ClSO$_2$–C$_6$H$_4$–NHCOCH$_3$ $\xrightarrow{\text{NaN}_3, \text{ acetone}}$ N$_3$SO$_2$–C$_6$H$_4$–NHCOCH$_3$

B. CH$_3$COCH$_2$CO$_2$Et $\xrightarrow{\text{N}_3\text{SO}_2\text{-C}_6\text{H}_4\text{-NHCOCH}_3}_{\text{NEt}_3, \text{CH}_3\text{CN}}$ CH$_3$CO-C(N$_2$)-CO$_2$Et

C. CH$_3$CO-C(N$_2$)-CO$_2$Et + PhC≡CH $\xrightarrow{\text{Rh}_2(\text{OAc})_2}_{\text{CH}_2\text{Cl}_2}$ 2-methyl-5-phenyl-3-furancarboxylic acid ethyl ester

Submitted by Huw M. L. Davies,[1] William R. Cantrell, Jr.,[1] Karen R. Romines,[1] and Jonathan S. Baum.[2]
Checked by Frank Stappenbeck and James D. White.

1. Procedure

Caution! These reactions, which involve toxic reagents, should be carried out in an efficient hood. Although p-acetamidobenzenesulfonyl azide exhibited no impact sensitivity,[3] proper caution should be exercised with all azide compounds.

A. *p-Acetamidobenzenesulfonyl azide.*[3] A 2-L Erlenmeyer flask equipped with a magnetic stirrer is charged with 117.0 g (0.50 mol) of p-acetamidobenzenesulfonyl chloride (Note 1) and 1 L of acetone. A solution of 39.0 g (0.60 mol) of sodium azide in 300 mL of water is added with stirring and the resulting mixture is left to stir for 12 hr. Three 2-L beakers equipped with magnetic stirrers are charged with 1.5 L each of water. The reaction mixture is divided into three portions and poured into the beakers with stirring. After the mixture is stirred for 1 hr, the white precipitate is filtered (Note 2) and dried in a desiccator over sodium hydroxide for 24 hr. Recrystallization of this material in four portions from toluene (1.5 L each portion), while the temperature is maintained below 80°C (Note 3), affords 88.9 g (74%) of p-acetamidobenzenesulfonyl azide as white crystals, mp 113°-115°C (Note 4).

B. *Ethyl diazoacetoacetate.*[3] A 2-L, round-bottomed flask equipped with a magnetic stirrer is charged with 26.0 g (0.20 mol) of ethyl acetoacetate, 49.0 g, (0.20 mol) of p-acetamidobenzenesulfonyl azide and 1.5 L of acetonitrile. The reaction vessel is cooled in an ice bath, and 60.6 g (0.60 mol) of triethylamine is added to the stirring mixture in one portion. The reaction mixture is warmed to room temperature and stirred for 12 hr. The solvent is removed under reduced pressure, and the residue is triturated with 500 mL of a 1:1 mixture of ether/petroleum ether. The mixture is filtered to remove the sulfonamide by-product, and the filtrate and wash are concentrated under reduced pressure. The crude product is purified by chromatograpy on silica gel (130 g, Note 5) with ether/petroleum ether (1:4) as eluant to yield 28.5 g (91%) of ethyl diazoacetoacetate as a yellow oil (Note 6).

C. *Ethyl 2-methyl-5-phenyl-3-furancarboxylate.*[4] A 1-L, three-necked, round-bottomed flask equipped with a magnetic stirrer, an addition funnel, and a reflux condenser is flushed with argon (Note 7). The reaction vessel is charged with 44.35 g of phenylacetylene (0.44 mol, Note 8), 0.38 g of rhodium(II) acetate dimer (0.00087 mol), and 100 mL of dichloromethane and the mixture is heated to reflux under an argon atmosphere. The addition funnel is charged with 13.57 g of ethyl diazoacetoacetate (0.087 mol) and 200 mL of dichloromethane, and this solution is added dropwise over 1.5 hr to the reaction mixture. After the reaction mixture is heated under reflux for an additional 12 hr, it is cooled and the solvent is removed under reduced pressure. The crude product is purified by chromatography on silica gel (110 g) with ether/petroleum ether (1:19) as eluant, followed by vacuum distillation (10-cm Vigreux column, 130°C, 0.1 mm) to yield 9.95 g (50%) of the furan as a pale yellow liquid (Note 9).

2. Notes

1. The following chemicals were obtained from the Aldrich Chemical Company, Inc., and were used without further purification: p-acetamidobenzenesulfonyl chloride, 97%; acetone, 99.9+%, HPLC grade; sodium azide, 99%; ethyl acetoacetate, 99%; triethylamine, 99%; rhodium(II) acetate dimer; phenylacetylene, 98%. The following solvents were obtained from Fisher Scientific and were used without further purification: toluene, certified A. C. S.; ethyl ether (Solvent grade, Concentrated); petroleum ether, certified A. C. S. Dichloromethane was distilled from calcium hydride.

2. The filtrate contains excess sodium azide which should be destroyed prior to disposal.

3. The azide partially decomposes at temperatures exceeding 80°C, and the resulting crystals appear slightly brown.

4. Data for p-acetamidobenzenesulfonyl azide are as follows: R_f = 0.49 (ether/petroleum ether (1:4)); ^{1}H NMR (200 MHz, CDCl$_3$) δ: 2.23 (s, 3 H), 7.75-7.89 (m, 4 H), 8.02 (s, 1 H); IR (nujol) cm^{-1}: 3250, 2110, 1665, 1580. Anal. Calcd for $C_8H_8N_4O_3S$: C, 40.00; H, 3.36; N, 23.32; S, 13.35. Found: C, 40.10; H, 3.40; N, 23.26; S, 13.40.

5. Silica gel 60 230-400 mesh ASTM was used. Whatman 250-mm layer, UV254, silica gel TLC plates with polyester backing were used to analyze the fractions.

6. Spectral data for ethyl diazoacetoacetate are as follows: ^{1}H NMR (200 MHz, CDCl$_3$) δ: 1.29 (t, 3 H, J = 7.1), 2.43 (s, 3 H), 4.26 (q, 2 H, J = 7.1); IR (neat) cm^{-1}: 2970, 2130, 1700, 1650.

7. The glassware in this reaction is dried with a heat gun and placed in a drying oven for 1 hr prior to use.

8. A smaller amount of phenylacetylene results in inefficient capture of the carbenoid intermediate, leading to lower yields.

9. Data for ethyl 2-methyl-5-phenyl-3-furancarboxylate are as follows: R_f = 0.51 (ether/petroleum ether (1:9)); ^{1}H NMR (200 MHz, CDCl$_3$) δ: 1.35 (t, 3 H, J = 7.1), 2.63 (s, 3 H), 4.30 (q, 2 H, J = 7.1), 6.87 (s, 1 H), 7.24-7.40 (m, 3 H), 7.60-7.65 (m, 2 H); IR (neat) cm^{-1}: 3080, 3000, 1725, 1610, 1590, 1565. Anal. Calcd for C$_{14}$H$_{14}$O$_3$: C, 73.03; H, 6.13. Found: C, 73.17; H, 6.09.

Waste Disposal Information

Excess sodium azide in the filtrate was destroyed by treatment with ammonium cerium(IV) nitrate solution according to the procedure described by Lunn, G.; Sansone, E. B., In "Destruction of Hazardous Chemicals in the Laboratory"; Wiley: New York, 1990; p. 44.

3. Discussion

The procedure described here provides a direct synthesis of highly substituted furans (see Table). Reaction of keto carbenoids with acetylenes is normally an efficient method to prepare cyclopropenes.[5] In numerous systems, however, the formation of furans was observed as a competing side reaction.[6] Furan formation is particularly favored when the carbenoid is a pyruvate[7] or contains two electron-withdrawing groups,[4,8] and when electron-donating groups are present on the acetylene.[4,8]

The diazo transfer reaction with sulfonyl azides has been used extensively for the preparation of diazo compounds.[9] Toluenesulfonyl azide is the standard reagent used,[10] but because of safety problems resulting from its potentially explosive nature, and because of the difficulty of product separation, several alternative reagents have been developed recently. n-Dodecylbenzenesulfonyl azide[11] has been reported to be very effective for the preparation of crystalline diazo compounds, while p-naphthalenesulfonyl azide[11] has been used for fairly non-polar compounds. Other useful reagents are methanesulfonyl azide[12] and p-carboxybenzenesulfonyl azide.[13] p-Acetamidobenzenesulfonyl azide[3] offers the advantages of low cost, safety, and ease of removal of the sulfonamide by-product through a simple trituration.

1. Department of Chemistry, Wake Forest University, P. O. Box 7486, Winston-Salem, NC 27109.
2. Agricultural Chemical Group, FMC Corporation, P. O. Box 8, Princeton, NJ 08543.
3. Baum, J. S.; Shook, D. A.; Davies, H. M. L.; Smith, H. D. *Synth. Commun.* **1987**, *17*, 1709.
4. Davies, H. M. L.; Romines, K. R. *Tetrahedron* **1988**, *44*, 3343.

5. Petiniot, N.; Anciaux, A. J.; Noels, A. F.; Hubert, A. J.; Teyssié, Ph. *Tetrahedron Lett.* **1978**, 1239.

6. (a) D'yakanov, I. A.; Komendantov, M. I. *J. Gen. Chem. USSR (Engl. Trans.)* **1961**, *31*, 3618; *Chem. Abstr.* **1962**, *57*, 8405f; (b) D'yakanov, I. A.; Komendantov, M. I.; Korshunov, S. P; *J. Gen. Chem. USSR (Engl. Trans.)* **1962**, *32*, 912; *Chem. Abstr.* **1963**, *58*, 2375e; (c) Komendantov, M. I.; Domnin, I. N.; Bulucheva, E. V. *Tetrahedron* **1975**, *31*, 2495; (d) Hendrick, M. E. *J. Am. Chem. Soc.* **1971**, *93*, 6337.

7. Wenkert, E.; Alonso, M. E.; Buckwalter, B. L.; Sanchez, E. L., *J. Am. Chem. Soc.* **1983**, *105*, 2021.

8. Yoshida, J.-i.; Yano, S.; Ozawa, T.; Kawabata, N. *J. Org. Chem.* **1985**, *50*, 3467.

9. Regitz, M., In "The Chemistry of Diazonium and Diazo Groups"; Patai, S., Ed.; Wiley: New York, 1978; Part 2, p. 751.

10. Regitz, M.; Hocker, J.; Liedhegener, A. *Org. Synth., Coll. Vol. V* **1973**, 179.

11. Hazen, G. G.; Weinstock, L. M.; Connell, R.; Bollinger, F. W. *Synth. Commun.* **1981**, *11*, 947.

12. Taber, D. F.; Ruckle, Jr., R. E.; Hennessy, M. J. *J. Org. Chem.* **1986**, *51*, 4077.

13. Hendrickson, J. B.; Wolf, W. A. *J. Org. Chem.* **1968**, *33*, 3610.

TABLE

RHODIUM(II) ACETATE-CATALYZED REACTION OF ACETYLENES WITH DIAZO CARBONYL COMPOUNDS[4]

Acetylene	Diazo	Furan	Yield %
Ph−C≡CH	EtO₂C−C(N₂)−C(O)CH₃	2-methyl-5-phenyl-3-(ethoxycarbonyl)furan	50
4-MeO-C₆H₄−C≡CH	CH₃C(O)−C(N₂)−C(O)CH₃	2-methyl-5-(4-methoxyphenyl)-3-acetylfuran	52
3,4-(MeO)₂-C₆H₃−C≡CH	EtO₂C−C(N₂)−C(O)CH₃	2-methyl-5-(3,4-dimethoxyphenyl)-3-(ethoxycarbonyl)furan	69

Appendix

Chemical Abstracts Nomenclature (Collective Index Number); (Registry Number)

Rhodium(II) acetate dimer: Acetic acid, rhodium(2+) salt (8,9); (5503-41-3)

Ethyl 2-methyl-5-phenyl-3-furancarboxylate: 3-Furoic acid, 2-methyl-5-phenyl-, ethyl ester (8); 3-Furancarboxylic acid, 2-methyl-5-phenyl-, ethyl ester (9); (29113-64-2)

p-Acetamidobenzenesulfonyl azide: Sulfanilyl azide, N-acetyl- (8); Benzenesulfonyl azide, 4-(acetylamino)- (9); (2158-14-7)

p-Acetamidobenzenesulfonyl chloride: Sulfanilyl chloride, N-acetyl- (8); Benzenesulfonyl chloride, 4-(acetylamino)- (9); (121-60-8)

Sodium azide (8,9); (26628-22-8)

Ethyl diazoacetoacetate: Acetoacetic acid, 2-diazo-, ethyl ester (8); Butanoic acid, 2-diazo-3-oxo-, ethyl ester (9); (2009-97-4)

Ethyl acetoacetate: Acetoacetic acid, ethyl ester (8); Butanoic acid, 3-oxo-, ethyl ester (9); (141-97-9)

IODOLACTAMIZATION:

8-exo-IODO-2-AZABICYCLO[3.3.0]OCTAN-3-ONE

(Cyclopenta[b]pyrrol-2(1H)-one, hexahydro-6-iodo-, (3aα,6α,6aα)-)

A. [cyclopentenyl-CH$_2$-COOH] $\xrightarrow{\text{1) (COCl)}_2\text{, toluene} \atop \text{2) NH}_3}$ [cyclopentenyl-CH$_2$-C(O)NH$_2$]

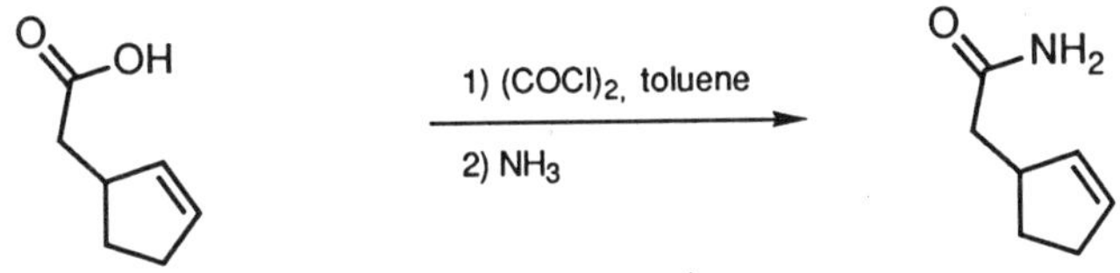

B. [cyclopentenyl-CH$_2$-C(O)NH$_2$] $\xrightarrow{\text{Me}_3\text{Si-OTf} \atop \text{Et}_3\text{N, pentane}}$ [Me$_3$Si–O–C(=N-SiMe$_3$)–CH$_2$–cyclopentenyl] $\xrightarrow{\text{1) I}_2\text{, ether} \atop \text{2) aq Na}_2\text{SO}_3\text{, aq Na}_2\text{CO}_3}$ [bicyclic lactam with I]

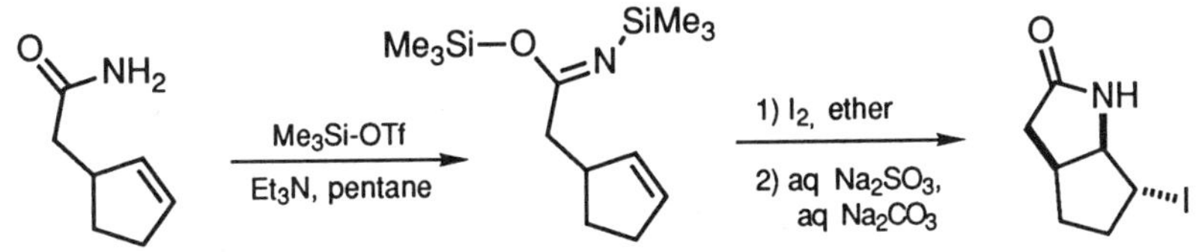

Submitted by Spencer Knapp and Frank S. Gibson.[1]
Checked by Chris Melville and James D. White.

1. Procedure

Caution! The following operations produce lachrymatory and corrosive vapors and must be carried out in a well-ventilated fume hood.

A. *2-Cyclopentene-1-acetamide.* A dry, 250-mL, three-necked, round-bottomed flask is equipped with a magnetic stirring bar, serum stopper, 25-mL pressure equalizing addition funnel, and an argon atmosphere with provision for venting gaseous reaction products (Note 1). The vessel is charged with 20 g (152 mmol) of 2-cyclopentene-1-acetic acid (Note 2) and 25 mL of dry toluene (Note 3). Oxalyl chloride (17.3 mL, 1.3 equiv) is added slowly over a 30-min period by means of the addition funnel, taking care to release any pressure buildup. (*Caution: gaseous hydrogen chloride evolution!*). The dark reaction mixture is stirred for an additional 20 min while the second reaction vessel is assembled, then concentrated to about 2/3 volume at the vacuum pump (Note 4).

A 250-mL, three-necked, round-bottomed flask equipped with magnetic stirring bar, serum stopper, dry ice condenser, and argon atmosphere is cooled by means of a dry ice/acetone bath and charged with approximately 150 mL of dry liquid ammonia. The 2-(2-cyclopentenyl)acetyl chloride reaction mixture is added carefully but steadily (Note 5) to the cold and rapidly stirred ammonia by syringe. (*Caution: vigorous exothermic reaction!*). Residual acid chloride is transferred by rinsing the first vessel with 5 mL of toluene. After the addition the mixture is stirred for an additional 5 min, then 60 mL of dichloromethane is added. The cold bath is removed and the excess ammonia is allowed to escape into the fume hood by stirring the open vessel overnight.

The crude reaction mixture is filtered and the filtrate is reserved. The solids are triturated by stirring vigorously with 100 mL of methanol for 20 min with gentle warming to about 40°C. The solids are filtered and triturated again in the same way. The three organic filtrates are combined and concentrated to near dryness. The resulting semi-solid is redissolved in 100 mL of dichloromethane, filtered to remove residual ammonium chloride, and concentrated to a light brown solid, 18.9 g. The crude product is dissolved in 60 mL of boiling tetrahydrofuran and allowed to crystallize in a -6°C freezer overnight. The amide is collected by filtration, washed with 5 mL of cold ether, and dried under reduced pressure, giving 15.77 g of white flakes, mp 128-129°C. The filtrate is concentrated to about 8 mL, brought to the cloud point by the addition of a few drops of hexane, and cooled in the freezer. Filtration as before gives a second crop of white flakes, 1.46 g, mp 128-129°C (total yield 17.23 g, 90.6%) (Note 6).

B. 8-exo-Iodo-2-azabicyclo[3.3.0]octan-3-one. A dry, 500-mL, three-necked, round-bottomed flask equipped with magnetic stirring bar, serum stopper, 50-mL pressure-equalizing addition funnel, cold water bath, and argon atmosphere is charged with 12.5 g (100 mmol) of 2-cyclopentene-1-acetamide, 29.2 mL (210 mmol) of triethylamine (Note 7), and 80 mL of dry pentane (Note 8). By means of the addition funnel, 41 mL (210 mmol) of trimethylsilyl trifluoromethanesulfonate (Note 9) is slowly added to the cooled and rapidly stirred amide suspension over a 50-min period. After the addition is complete, the reaction mixture is stirred for an additional 20 min at room temperature, then the stirring is stopped, and the two layers are allowed to separate.

A second, dry, 500-mL, three-necked, round-bottomed flask is equipped with magnetic stirring bar, serum stopper, vacuum pump connection, and argon atmosphere. The (top) pentane layer from the first flask, which contains the bis(trimethylsilyl)imidate, is carefully transferred to the second flask by cannula, maintaining the argon atmosphere, and leaving the oily triethylammonium trifluoromethanesulfonate layer behind. This remaining salt is triturated with 30 mL of a dry 2:1 pentane/ether mixture by stirring for 15 min, allowing the layers to separate, then transferring the extract to the second flask as before. The trituration is repeated with a 30-mL portion of anhydrous ether, and the combined extracts in the second flask are concentrated with stirring to about 1/3 volume using the vacuum pump (Note 10).

A third, dry, 500-mL, three-necked, round-bottomed flask equipped with an addition funnel, magnetic stirring bar, serum stopper, cold water bath, and argon atmosphere is charged with 53.3 g (210 mmol) of molecular iodine and 140 mL of anhydrous ether. The mixture is allowed to stir for 10 min to dissolve most of the iodine. The concentrated organic extract in the second flask is now added to the iodine solution with stirring and cooling over 15 min. The reaction mixture warms slightly during the addition, but should not reach reflux. An additional 10 mL of anhydrous ether is used to complete the transfer of the bis(trimethylsilyl)imidate. Near the end of the addition, the oily black layer (which contains the cyclized iminium salt) solidifies, leaving a clear light brown ether supernatant. The serum stopper is carefully removed and the solid residue gently broken up using a spatula. The stopper is replaced and the reaction mixture is allowed to stand for an additional 45 min with occasional swirling by hand. The reaction is quenched by removing the addition funnel and stopper and slowly adding 20 mL of saturated aqueous sodium carbonate. (*Caution: vigorous gas evolution!*). At this point stirring can be resumed. A 20-mL portion of saturated aqueous sodium sulfite is added slowly (*more gas evolution*), and the process is repeated until 100 mL each of saturated aqueous sodium carbonate and sulfite have been added. The reaction mixture is filtered, the crude solid iodolactam is reserved, and the organic layer is separated and reserved. The aqueous layer is saturated with sodium chloride and extracted with four, 100-mL portions of dichloromethane. The five organic extracts are combined, dried over anhydrous sodium sulfate, and concentrated to a light brown solid. This residue is dissolved in 20 mL of tetrahydrofuran and hexane is added to the cloud point. Cooling in the freezer gives colorless needles, which are collected and dried under reduced pressure to afford 1.60 g of iodolactam, mp 138-139°C.

The reserved solid from filtration is dried under reduced pressure, dissolved in 65 mL of hot tetrahydrofuran and filtered. The solution is allowed to cool, first at room temperature, then in the freezer. The product (16.77 g, mp 138-139°C) is collected as before. The mother liquor is concentrated to about 10 mL, brought to the cloud point by the addition of hexane, and cooled in the freezer, resulting in an additional crop of 1.70 g, mp 138-139°C. The total amount of iodolactam is 19.80 g, representing a 79% yield from the amide (Note 11).

2. Notes

1. All reaction glassware was oven dried at 120°C and assembled hot. The submitters used three evacuate/fill cycles from an argon-filled balloon fitted on a three-way stopcock to provide the inert atmosphere.

2. 2-Cyclopentene-1-acetic acid (96%) was purchased from Aldrich Chemical Company, Inc., and used as received.

3. Toluene was dried by distillation from -40 mesh calcium hydride. Unless otherwise specified, reagents in this procedure were obtained commercially and used as received.

4. In series with the usual (500 mL) dry ice/acetone trap, a trap filled with solid sodium hydroxide was used to protect the pump from acidic vapors.

5. Continuous addition of the carboxylic acid chloride solution is required to prevent clogging the syringe needle.

6. The spectral properties are as follows: FT-IR (KBr) cm^{-1}: 3360, 3355, 1663, 1634; ^{1}H NMR (CDCl$_3$, 400 MHz) δ: 1.42-1.49 (m, 1 H), 2.07-2.35 (m, 5 H), 3.06-3.09 (m, 1 H), 5.66 (br s, 1 H), 5.75 (app dd, 1 H, J = 2, 5), 5.76 (app dd, 1 H, J = 2.5, 4.5), 5.94 (br s, 1 H); ^{13}C NMR δ: 29.6, 31.8, 42.0, 42.3, 131.7, 133.7, 175.0.

7. Triethylamine was dried by distillation from -40 mesh calcium hydride.

8. Pentane was dried by distillation from -40 mesh calcium hydride.

9. Trimethylsilyl trifluoromethanesulfonate (99%) was purchased from Aldrich Chemical Company, Inc., and used as received.

10. Any adventitious water introduced during these operations results in a decreased yield of iodo lactam and the formation of iodo lactone as an undesired side product.

11. The spectral properties are as follows: FT-IR (KBr) cm^{-1}: 3189, 3078, 3034, 1680; ^{1}H NMR (CDCl$_3$, 400 MHz) δ: 1.55-1.60 (m, 1 H), 1.99-2.06 (m, 1 H), 2.10-2.20 (m, 2 H), 2.33-2.45 (m, 1 H), 2.70 (dd, 1 H, J = 18, 10), 3.05-3.13 (m, 1 H), 4.16 (br s, 1 H), 4.40 (d, 1 H, J = 7.2), 5.75-5.85 (br s, 1 H); ^{13}C NMR δ: 32.3, 32.9, 35.0, 35.5, 38.2, 69.8, 178.6.

Waste Disposal Information

All toxic materials were disposed of in accordance with "Prudent Practices for Disposal of Chemicals from Laboratories"; National Academy Press; Washington, DC, 1983.

3. Discussion

This "iodolactamization" procedure has been optimized for the present example. Related reaction conditions have been used to generate a series of iodo lactams from the corresponding unsaturated amides (Table).[2] For these (smaller scale) examples, the cyclization was carried out in tetrahydrofuran solution, and isolation was by column chromatography. The lactams in entries 3, 4, and 8 have also been recently prepared by the submitters in 81%, 82%, and 78% yields, respectively, using the procedure described here. Cyclization in ether solution rather than in tetrahydrofuran avoids the formation of iodobutanol,[2] which must then be separated from product.

Iodo lactams are a useful, new class of difunctional compounds. Conversions of iodo lactams to N-acylaziridines,[2,3] unsaturated lactams,[2,4] azido lactams,[2,3] amino lactams,[2,3] hydroxy lactams,[2] annulated lactams,[2,5] and other derivatives have been described. Halo lactams have also been prepared from aspartic acid[5] and glutamic acid,[6] from N-substituted unsaturated amides[7-9] and imidates,[10-12] and from unsaturated amides whose competing O-cyclization reaction is less favored.[13,14]

1. Department of Chemistry, Rutgers The State University of New Jersey. New Brunswick, NJ 08903.
2. Knapp, S.; Levorse, A. T. *J. Org. Chem.* **1988**, *53*, 4006, and references therein. Preliminary communication: Knapp, S.; Rodriques, K. E.; Levorse, A. T.; Ornaf, R. M. *Tetrahedron Lett.* **1985**, *26*, 1803.
3. Knapp, S.; Levorse, A. T. *Tetrahedron Lett.* **1987**, *28*, 3213.
4. Knapp, S.; Levorse, A. T.; Potenza, A. J. *J. Org. Chem.* **1988**, *53*, 4773.
5. Salzmann, T. N.; Ratcliffe, R. W.; Christensen, B. G.; Bouffard, F. A. *J. Am. Chem. Soc.* **1980**, *102*, 6161.
6. Hardegger, E.; Ott, H. *Helv. Chim. Acta* **1955**, *38*, 312.
7. Biloski, A. J.; Wood, R. D.; Ganem, B. *J. Am. Chem. Soc.* **1982**, *104*, 3233.
8. (a) Rajendra, G.; Miller, M. J. *Tetrahedron Lett.* **1985**, *26*, 5385; (b) Rajendra, G.; Miller, M. J. *J. Org. Chem.* **1987**, *52*, 4471; (c) Rajendra, G.; Miller, M. J. *Tetrahedron Lett.* **1987**, *28*, 6257.
9. Balko, T. W.; Brinkmeyer, R. S.; Terando, N. H. *Tetrahedron Lett.* **1989**, *30*, 2045.
10. Kano, S.; Yokomatsu, T.; Iwasawa, H.; Shibuya, S. *Heterocycles* **1987**, *26*, 359.
11. Takahata, H.; Takamatsu, T.; Mozumi, M.; Chen. Y.-S.; Yamazaki, T.; Aoe, K. *J. Chem. Soc., Chem. Commun.* **1987**, 1627.
12. Kurth, M. J.; Bloom, S. H. *J. Org. Chem.* **1989**, *54*, 411.
13. Aida, T.; Legault, R.; Dugat, D.; Durst, T. *Tetrahedron Lett.* **1979**, 4993.
14. (a) Tamaru, Y.; Kawamura, S.; Tanaka, K.; Yoshida, Z. *Tetrahedron Lett.* **1984**, *25*, 1063; (b) Tamaru, Y.; Kawamura, S.-i. Bando, T.; Tanaka, K.; Hojo, M.; Yoshida, Z.-i. *J. Org. Chem.* **1988**, *53*, 5491.

TABLE

PREPARATION OF IODO LACTAMS

Entry	Unsaturated Amide	Iodo Lactam(s)	% Yield (cis/trans)
1	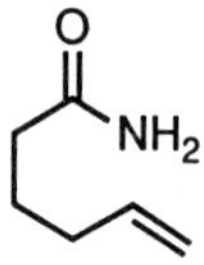 CH$_2$=CHCH$_2$CH$_2$CH$_2$C(O)NH$_2$ (pent-4-enamide)	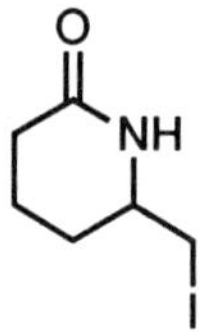 6-(iodomethyl)piperidin-2-one	35[a]
2	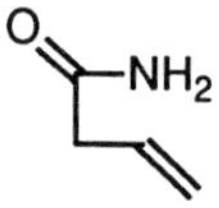 CH$_2$=CHCH$_2$C(O)NH$_2$ (but-3-enamide)	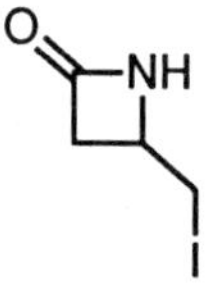 4-(iodomethyl)azetidin-2-one	35[b]
3	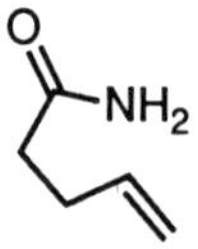 CH$_2$=CHCH$_2$CH$_2$C(O)NH$_2$ (pent-4-enamide shorter)	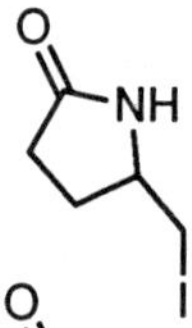 5-(iodomethyl)pyrrolidin-2-one	86
4	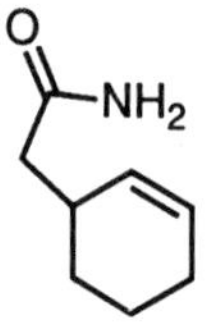 2-(cyclohex-2-enyl)acetamide	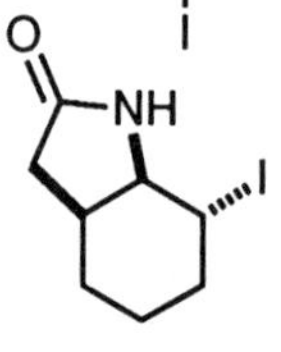 iodo-octahydroindol-2-one	88
5	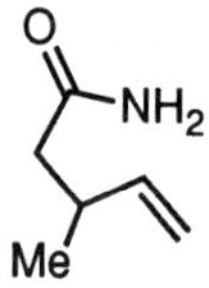 3-methylpent-4-enamide	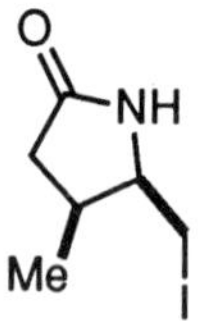 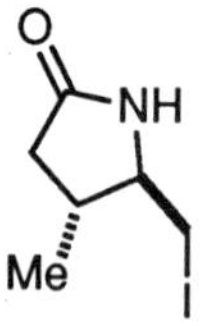 cis- and trans-4-methyl-5-(iodomethyl)pyrrolidin-2-one	84 (3:1)

Entry	Unsaturated Amide	Iodo Lactam(s)	% Yield (cis/trans)
6	CH₂=CH-CH(Ph)-CH₂-C(O)-NH₂	5-(iodomethyl)-4-phenyl-pyrrolidin-2-one (cis) + (trans)	88 (2:1)
7	CH₂=CH-CH(CH₂OCH₂Ph)-CH₂-C(O)-NH₂	5-(iodomethyl)-4-(benzyloxy)-pyrrolidin-2-one (cis) + (trans)	86 (1:1)
8	norbornene-2-carboxamide	iodo-bicyclic lactam	80[c]

[a] 10-20% of starting amide was also recovered. [b] 54% of crotonamide was also isolated. [c] Overall yield after separate desilylation.

Appendix

Chemical Abstracts Nomenclature (Collective Index Number);

(Registry Number)

8-exo-Iodo-2-azabicyclo[3.3.0]octan-3-one: Cyclopenta[b]pyrrol-2(1H)-one, hexahydro-6-iodo-, (3aα,6α,6aα)- (11); (100556-58-9)

2-Cyclopentene-1-acetamide (10); (72845-09-1)

2-Cyclopentene-1-acetic acid (8,9); (13668-61-6)

Oxalyl chloride (8); Ethanedioyl dichloride (9); (79-37-8)

Trimethylsilyl trifluoromethanesulfonate: Methanesulfonic acid, trifluoro-, trimethylsilyl ester (8,9); (27607-77-8)

(E)-1-BENZYL-3-(1-IODOETHYLIDENE)PIPERIDINE: NUCLEOPHILE-PROMOTED ALKYNE-IMINIUM ION CYCLIZATIONS

A. $CH_3-\equiv-\sim-OH$ + CH_3SO_2Cl $\xrightarrow{Et_3N}$ $CH_3-\equiv-\sim-OSO_2CH_3$
1

B. $CH_3-\equiv-\sim-OSO_2CH_3$ + $PhCH_2NH_2$ $\xrightarrow[Me_2SO]{NaI (cat.)}$ $CH_3-\equiv-\sim-NH-CH_2Ph$
2

C. $CH_3-\equiv-\sim-N(CH_2Ph)H$ + HCHO $\xrightarrow[CampSO_3H]{NaI}$ **3**

Submitted by H. Arnold, L. E. Overman, M. J. Sharp and M. C. Witschel.[1]
Checked by Antje Grützmann, Thomas Hache, and Ekkehard Winterfeldt.

1. Procedure

A. 4-Hexyn-1-yl methanesulfonate (**1**). An oven-dried, 500-mL, one-necked, round-bottomed flask equipped with a magnetic stirring bar is flushed with argon and 260 mL of dichloromethane is added (Note 1). The flask is sealed with a rubber septum inlet and cooled to ca. -10°C in an ice-salt bath. To this flask are added via syringe 11 mL (80 mmol) of triethylamine, 5.0 g (51 mmol) of 4-hexyn-1-ol (Note 2), and 4.3 mL (56 mmol) of methanesulfonyl chloride (Note 3). The resulting solution is stirred for an additional 30 min and then quenched by adding 30 mL of ice-water. The organic layer is separated and washed successively with 1 M hydrochloric acid solution (30 mL), saturated aqueous sodium bicarbonate solution (30 mL), and brine (30 mL). The organic layer is dried over magnesium sulfate, filtered, and concentrated with a rotary evaporator to give 8.3-9.0 g (93-100%) of crude 4-hexyn-1-yl methanesulfonate (**1**) which was used directly in the next step (Note 4).

    B. N-Benzyl-4-hexyn-1-amine (**2**). An oven-dried, 100-mL, one-necked, round-bottomed flask containing a magnetic stirring bar and a rubber septum inlet is flushed with argon and charged with 200 mg of sodium iodide. The crude mesylate **1**, 40 mL of dimethyl sulfoxide (Note 5) and 10.9 g (102 mmol) of benzylamine are added via syringe. The resulting solution is heated in an oil bath at 47-53°C for 5 hr (Note 6) and then allowed to cool to room temperature. The reaction solution is poured into a separatory funnel containing 200 mL of aqueous 1% sodium hydroxide solution and the resulting mixture is extracted with ether (3 x 100 mL). The combined ether extracts are washed with brine (50 mL), dried over magnesium sulfate, filtered, and concentrated with a rotary evaporator. The residue (ca. 9 g) is purified by flash chromatography (ca. 300 g of silica gel using 1:1 hexane-ether containing 5% triethylamine as the eluent) (Note 7) to give 7.2-7.5 g (75-79% overall) of **2** as a colorless liquid (Note 8).

    C. (E)-1-Benzyl-3-(1-iodoethylidene)piperidine (**3**). A 250-mL, one-necked, round-bottomed flask containing a magnetic stirring bar and a reflux condenser topped with a rubber septum inlet is flushed with argon and charged with 4.0 g (21 mmol) of alkynylamine **2**, 11 g (73 mmol) of sodium iodide (Note 9), 35 mL of 37% w/w formaldehyde solution, 5.4 g (22 mmol) of camphorsulfonic acid monohydrate (Note 10) and 80 mL of water. The resulting mixture is heated at reflux under an argon atmosphere for 15 min (Note 11) and then allowed to cool to room temperature. This solution is made basic by adding 5 M aqueous potassium hydroxide solution and then poured into a separatory funnel where it is extracted with dichloromethane (3 x 50 mL) (Note 12). The combined organic layers are dried over sodium sulfate, filtered, and concentrated with a rotary evaporator. The resulting residue is purified by flash chromatography (ca. 150 g of silica gel, 1:1 hexane-ethyl ether containing 5% triethylamine as eluent) to give 5.4-6.2 g (79-90%) of **3** as a colorless oil (Notes 13 and 14).

2. Notes

1. Dichloromethane is distilled from calcium hydride (CaH_2) and added directly from the still to the reaction flask.

2. This alcohol is readily prepared in standard fashion from the tetrahydropyranyl ether of 4-pentyn-1-ol and iodomethane: A hexane solution of butyllithium (46 mL of a 2.2 M solution, 100 mmol) is added dropwise under an argon atmosphere to a dry ice-cooled solution of tetrahydro-2-(4-pentynyloxy)-2H-pyran (14.4 g, 84 mmol, prepared from commercially available 4-pentyn-1-ol[2]) in 100 mL of dry tetrahydrofuran. After 10 min, 7.0 mL (110 mmol) of iodomethane is added dropwise to the dry ice-cooled, stirring solution of the alkynyllithium intermediate. The reaction mixture is maintained at dry ice temperature for 1 hr, and, after warming to room temperature, the tetrahydrofuran is removed by mild rotary evaporation (or distillation at atmospheric pressure). The crude product is dissolved in 100 mL of ether and washed with brine (50 mL), and the aqueous phase is back-extracted with ether (25 mL). The combined organic phases are concentrated and the residue is dissolved in 200 mL of methanol. p-Toluenesulfonic acid (4 g) is added and the resulting solution is heated at reflux for 3 hr. The solvent is removed by distillation through an 8-10 cm Vigreux column, the residue is dissolved in 100 mL of ether and this solution is extracted with 25 mL of aqueous 10% sodium carbonate solution. After

the solution is dried over MgSO$_4$, the solvent is removed by distillation and the residue is distilled through a 50-cm concentric tube column. The fraction boiling at 92-93°C (10 mm Hg) is collected to give 7.8 g (94%) of 4-hexyn-1-ol, which is >95% pure by GC analysis (30 m, Supelco SPB-5 capillary column).

3. Triethylamine, distilled from CaH$_2$, and methanesulfonyl chloride, vacuum distilled at ca. 60°C (20 mm), are employed.

4. The spectrum is as follows: ^{1}H NMR (300 MHz, CDCl$_3$) δ: 1.78 (t, 3 H, J = 2.5), 1.91 (apparent pentaplet, 2 H, J = 6.5), 2.25-2.35 (m, 2 H), 3.03 (s, 3 H), 4.35 (t, 2 H, J = 6.1).

5. Dimethyl sulfoxide, distilled at 20 mm from CaH$_2$, and benzylamine, freshly distilled at 20 mm, are employed.

6. The reaction is easily monitored by TLC (silica gel, 1:1 ether-ethyl acetate): mesylate R_f = 0.9, amine R_f = 0.4.

7. This material can also be purified by vacuum distillation; however, some decomposition results.

8. This material is 97% pure by capillary GC analysis (30 m, J & W DB-5 fused silica column). Spectral data are as follows: IR (film) cm^{-1}: 3330, 1453, 1120, 736; ^{1}H NMR (300 MHz, CDCl$_3$) δ: 1.69 (apparent pentaplet, 2 H, J = 7), 1.77 (t, 3 H, J = 2.5), 2.15-2.25 (m, 2 H), 2.73 (t, 2 H, J = 7.0), 3.80 (s, 2 H), 7.2-7.4 (m, 5 H); Mass spectrum (isobutane CI): 188 (MH), 172, 120, 91; high resolution mass spectrum (70 eV, EI) 187.1331 (187.1261 Calcd for C$_{13}$H$_{17}$N).

9. Fisher *Certified* sodium iodide is used as received.

10. Fisher *Certified* A.C.S. formaldehyde solution is used as received. Aldrich Chemical Company, Inc., camphorsulfonic acid monohydrate is recrystallized from ethyl acetate prior to use.

11. This conversion is easily monitored by TLC (silica gel, 1:1 hexane-ethyl acetate containing 5% triethylamine): **2**, R_f = 0.4, **3**, R_f = 0.7.

12. The free base of this iodoamine darkens slowly when exposed to room light. The isolation procedure should be conducted rapidly or the separatory funnel and rotary evaporator bulb should be wrapped in aluminum foil to exclude room light.

13. This material is at least 95% pure by capillary GC analysis (30 m, J & W DB-5 fused silica column); a small unknown impurity (ca. 2%) with characteristic ^{1}H NMR signals at δ 3.41, 4.77 and 4.92 is apparent in some chromatography fractions. Spectral data for **3** are as follows: IR (film) cm^{-1}: 1646, 1228, 1138, 1119, 1061, 739; ^{1}H NMR δ: 1..55-1.8 (m, 2 H), 2.40 (t, 2 H, J = 6.3), 2.45 (s, 3 H), 2.56 (t, 2 H, J = 5.5), 3.10 (s, 2 H), 3.57 (s, 2 H), 7.2-7.4 (m, 5 H); mass spectrum (isobutane CI): 328 (MH), 202, 200, 112, 110, 92. High resolution mass spectrum (70 eV, EI): 327.0465 (327.0484 calcd for $C_{14}H_{18}NI$).

14. The maleate salt is prepared in good yield and crystallizes as fine needles (ca. 1 g of salt/10 mL) from absolute ethanol: mp 143-144°C. This salt can be stored at room temperature in room light with no noticeable decomposition. Anal. Calcd. for $C_{18}H_{22}INO_4$: C, 48.77; H, 5.00; N, 3.16. Found: C, 48.70; H, 5.00; N, 3.09.

Waste Disposal Information

All toxic materials were disposed of in accordance with "Prudent Practices for Disposal of Chemicals from Laboratories"; National Academy Press; Washington, DC, 1983.

3. Discussion

Simple alkynes do not undergo intramolecular reactions with weakly electrophilic iminium ions in the absence of strong external nucleophiles.[3] For example, the formaldiminium ion derived from 4-hexynylamine, formaldehyde, and camphorsulfonic acid does not cyclize when maintained in acetonitrile at 100°C for 1 hr.[3] Iminium ion-alkyne cyclizations do take place in nucleophilic solvents such as H_2O[4] or in non-nucleophilic solvents when a strong nucleophile is present.[3]

The present procedure illustrates the use of added iodide anion to promote the Mannich cyclization of an alkyne to afford 3-alkylidenepiperidines. As illustrated in Table I a variety of nonbasic nucleophiles with nucleophilic constants[5] $\eta\text{-}CH_3I$ >5.8 are useful promoters of formaldiminium ion-alkyne cyclizations.[3] Piperidines containing both endocyclic and exocyclic allylic unsaturation can be efficiently assembled in this way from readily available alkynol precursors (see Table I).[3] To the limits of ^1H NMR detection at 500 MHz all nucleophile-promoted cyclizations that form 3-alkylidenepiperidines occur with complete anti-stereoselectivity.

The usefulness of nucleophile-promoted iminium ion-alkyne cyclizations derives from the ready availability of alkynylamines and the subsequent transformations of the cyclization products made possible because of their vinylic functionality (e.g., equations 1 and 2). Equation 2 illustrates use of this chemistry to elaborate an exocyclic tetrasubstituted double bond with complete stereocontrol. Net "reductive" iminium ion alkyne cyclizations can be accomplished by dehalogenation of vinyl halide cyclization products. The conversion illustrated in equation 3 is a key step in an efficient, practical synthesis of the cardiotonic frog alkaloid pumiliotoxin A.[6]

TABLE I. Nucleophile-Promoted Iminium Ion-Alkyne Cyclizations[3]

R'	R	X	Yield
OCH$_3$	CH$_3$	Br	89%
OCH$_3$	CH$_3$	I	80%
OCH$_3$	n-Bu	Br	63%[7]
OCH$_3$	n-Bu	I	76%[7]
H	CH$_3$	I	81-90%
H	CH$_3$	N$_3$	72%
H	CH$_3$	SCN	82%
OCH$_3$	H	I	56%
OCH$_3$	CH$_3$	SPh	<15%

R	X	Yield
CH$_3$	Br	75%
CH$_3$	I	90%
CH$_3$	N$_3$	40%
Me$_3$Si	Br	44%
Me$_3$Si	I	60%
H	I	87%

(+)-pumiliotoxin A

1. Department of Chemistry, University of California, Irvine, CA 92717.
2. Negishi, E.-i.; Chiu, K.-W. *J. Org. Chem.* **1976**, *41*, 3484.
3. Overman, L. E.; Sharp, M. J. *J. Am. Chem. Soc.* **1988**, *110*, 612, 5934.
4. Ahmad, V. U.; Feuerherd, K. H.; Winterfeldt, E. *Chem. Ber.* **1977**, *110*, 3624.
5. Pearson, R. G.; Sobel, H.; Songstad, J. *J. Am. Chem. Soc.* **1968** *90*, 319.
6. Overman, L. E.; Sharp, M. J. *Tetrahedron Lett.* **1988**, *29*, 901.
7. Sharp, M. J. unpublished examples.

Appendix

Chemical Abstracts Nomenclature (Collective Index Number);
(Registry Number)

4-Hexyn-1-yl methanesulfonate: 4-Hexyn-1-ol, methanesulfonate (10); (68275-05-8)

4-Hexyn-1-ol (8,9); (928-93-8)

4-Pentyn-1-ol (8,9); (5390-04-5)

Tetrahydro-2-(4-pentynyloxy)-2H-pyran: 2H-Pyran, tetrahydro-2-(4-pentynyloxy)-
(10); (62992-46-5)

Iodomethane: Methane, iodo- (8,9); (74-88-4)

Methanesulfonyl chloride (8,9); (124-63-0)

N-Benzyl-4-hexyn-1-amine: Benzenemethanamine, N-4-hexynyl- (12); (112069-91-7)

Dimethyl sulfoxide: Methyl sulfoxide

(8); Methane, sulfinyl bis- (9); (67-68-5)

Benzylamine (8); Benzenemethanamine (9); (100-46-9)

Sodium iodide (8,9); (7681-82-5)

Formaldehyde (8,9); (50-00-0)

Camphorsulfonic acid monohydrate: Bicyclo[2.2.1]heptane-1-methanesulfonic acid,
7,7-dimethyl-2-oxo-, (±)- (9); (5872-08-2)

9-THIABICYCLO[3.3.1]NONANE-2,6-DIONE

A. SCl₂ + (cycloocta-1,5-diene) ⟶ 2,6-dichloro-9-thiabicyclo[3.3.1]nonane (**1**)

B. **1** $\xrightarrow{\text{Na}_2\text{CO}_3, \text{H}_2\text{O}}$ 2,6-dihydroxy-9-thiabicyclo[3.3.1]nonane (**2**)

C. **2** $\xrightarrow[\text{CH}_3\text{SOCH}_3]{(\text{COCl})_2}$ 9-thiabicyclo[3.3.1]nonane-2,6-dione (**3**)

Submitted by Roger Bishop.[1]
Checked by Graham N. Maw and Robert K. Boeckman, Jr.

1. Procedure

A. *(1α,2α,5α,6α)-2,6-Dichloro-9-thiabicyclo[3.3.1]nonane* (**1**). *(Caution! Preparation A should be carried out in a well-ventilated hood).* A dry, 2-L, four-necked, round-bottomed flask is equipped with a sealed mechanical stirrer (Note 1), 1-L pressure-equalizing funnel fitted with a drying tube, low temperature thermometer, and a nitrogen inlet. The flask is charged with 125 mL (1.02 mol) of 1,5-cyclooctadiene (Note 2) and 1 L of reagent dichloromethane, cooled to -50 to -60°C using an external acetone-dry ice bath, and the solution placed under a slow stream of dry nitrogen. To the vigorously stirred solution is added slowly over a period of 2 hr a solution of 65 mL (1.02 mol) of freshly purified sulfur dichloride (Note 3) in 500 mL of dichloromethane while maintaining the temperature at, or below, -50°C. The cloudy solution is allowed to warm to room temperature and filtered to remove a small amount of white solid. The filtrate is transferred to a 3-L separatory funnel, washed with brine (3 x 100 mL), and dried (Na_2SO_4). Solvent is removed from the filtrate under reduced pressure with a rotary evaporator to afford 201.4-210.8 g (94-98%) of ($1\alpha,2\alpha,5\alpha,6\alpha$)-2,6-dichloro-9-thiabicyclo[3.3.1]nonane (**1**) as a faintly yellow solid, mp 97.5-99°C (Note 4).

B. *(endo,endo)-9-Thiabicyclo[3.3.1]nonane-2,6-diol* (**2**). To a 1-L, round-bottomed flask, equipped with a magnetic stirring bar, is added 21.12 g (0.10 mol) of the dichloride (**1**), 100 mL of acetone, and a solution of 34.33 g (0.12 mol) of sodium carbonate decahydrate (Note 5) in 200 mL of water. A condenser is attached to the flask, the contents are stirred gently and then heated to reflux. After 1 hr at reflux (bath temperature ca. 85°C) the clear solution is allowed to cool to room temperature and the stirrer bar is removed (Note 6). Solvent is removed under reduced pressure using a rotary evaporator until the aqueous slurry of white solid is reduced to roughly 25 mL and then the solid is removed by suction filtration at room temperature. The dried, crude diol is heated with 300 mL of methanol and the hot solution filtered directly into a 500-mL, round-bottomed flask, thereby removing small quantities of solid impurity (Note 7). A rotary evaporator is used to concentrate the solution to a volume of ca. 40 mL, and the resulting slurry is filtered to yield 15.2-16.2 g (87-93%) of (endo,endo)-9-thiabicyclo[3.3.1]nonane-2,6-diol (**2**) as a white solid (Notes 8, 9).

C. *9-Thiabicyclo[3.3.1]nonane-2,6-dione* (**3**). *(Caution! Oxalyl chloride and dimethyl sulfoxide are reported to react explosively at room temperature. Preparation C should be carried out in a well-ventilated hood since a co-product of the reaction is dimethyl sulfide).* To a 1-L, three-necked flask equipped with a magnetic stirring bar, low temperature thermometer, dropping funnel protected from moisture by a drying tube, and a second drying tube, is added 13.5 mL (0.16 mol) of oxalyl chloride and 350 mL of dichloromethane. The solution is stirred and cooled to -78°C using an external acetone-dry ice bath. A solution of 22.0 mL (0.31 mol) of dry dimethyl sulfoxide (Note 10) in 75 mL of dichloromethane is added over 10 min ensuring that the reaction temperature does not exceed -60°C. After a further 10 min a solution of 13.07 g (0.08 mol) of (endo,endo)-9-thiabicyclo[3.3.1]nonane-2,6-diol (**2**) in 30 mL of dry dimethyl sulfoxide is added rapidly from the dropping funnel to the stirred solution at -78°C (Note 11). Final traces of the solution are washed into the reaction using a further 10 mL of dry dimethyl sulfoxide. The reaction is stirred for 25 min at -78°C after addition of diol **2** is complete and then 105 mL of redistilled triethylamine is added dropwise. After a further 15 min at -78°C the cooling bath is removed and the reaction is allowed to warm to room temperature whereupon 300 mL of water is added. The material is transferred to a 2-L separatory funnel, the dichloromethane layer is separated, and the aqueous layer is extracted with two 100-mL portions of dichloromethane. The combined dichloromethane extracts are washed successively with 1 L of 1% hydrochloric acid, 300 mL of 5% aqueous sodium carbonate, and two 300-mL portions of water. After the extracts are dried (anhydrous sodium sulfate), the pale yellow filtrate is evaporated to give 11.9-12.4 g of crude 9-thiabicyclo[3.3.1]nonane-2,6-dione (**3**) as a slightly yellow solid. Thin layer chromatography (silica/dichloromethane) indicates the presence of a small quantity of colored polar impurity, which is not easily removed by recrystallization. The crude solid is dissolved in ca. 300 mL of hot 1:1 ethyl ether/benzene containing 0.5 g of activated charcoal, then filtered through a short plug of Celite filter aid. The plug is washed with further hot solvent and the combined filtrates are evaporated to give 11.3-11.7 g (88-92%) of 9-thiabicyclo[3.3.1]nonane-2,6-dione (**3**) as a faintly off-white solid of ca. 99% purity. Analytically pure material is obtained by recrystallization from 2:1 light petroleum/chloroform or from 1:1 light petroleum/dichloromethane.

Alternatively, the product is sublimed under reduced pressure (ca. 160°C/30 mm) (Note 12).

2. Notes

1. A magnetic stirring bar may be used provided that *vigorous* mixing of the solution is possible.

2. 1,5-Cyclooctadiene purchased from the Aldrich Chemical Company, Inc. was purified by elution of the neat liquid through a short column of activated alumina immediately prior to use.

3. Sulfur dichloride partially decomposes on standing to chlorine and sulfur monochloride and so should always be purified before use. Commercial sulfur dichloride of approximately 80% purity (Aldrich Chemical Company, Inc.) was purified in a hood following the procedure described by Brauer.[2] Phosphorus trichloride (2 mL) was added to crude sulfur dichloride (200 mL) contained in a flask set up for a standard distillation and protected from the atmosphere by a silica gel drying tube. The fraction boiling between 55-63°C was collected in a receiving flask containing phosphorus trichloride (5 drops). A second distillation using the same technique gave pure sulfur dichloride of bp 59-60°C/atmospheric pressure. Material stabilized with a few drops of phosphorus trichloride may be stored for a few days in a sealed container without significant decomposition.

4. Compound **1** has the following spectral properties. IR (paraffin mull) cm^{-1}: 1245 (m), 1160 (m), 1000 (w), 950 (w), 890 (w), 815 (s), 755 (m), 735 (m); ^{1}H NMR (CDCl$_3$) δ: 2.18-2.38 (6 H, m), 2.65-2.70 (2 H, m), 2.85-2.88 (2 H, m), 4.68-4.75 (2 H, m); ^{13}C NMR (CDCl$_3$) δ: 28.3 (t), 32.6 (t), 37.3 (d), 62.5 (d). The crude product is sufficiently pure for most purposes, but analytically pure material is colorless, lit.[3] mp 98.1-99.6°C. This may be obtained by trituration of the crude product with a little ethyl

ether (which removes a small quantity of yellowish oil) followed by filtration. Alternatively it may be purified by sublimation under reduced pressure (48°C/0.05 mm)[3] or recrystallized from benzene.[4]

5. The checkers employed anhydrous sodium carbonate.

6. The checkers filtered the solution at this point to remove small amounts of a brown solid.

7. (endo,endo)-9-Thiabicyclo[3.3.1]nonane-2,6-diol (**2**) is remarkably insoluble in the solvents normally used for extraction of organic materials from aqueous solutions.

8. Compound (**2**) has the following spectral properties. IR (paraffin mull) cm^{-1}: 3300 (s), 1025 (s), 990 (m), 950 (w), 900 (w), 880 (m): ^{1}H NMR (d^6-DMSO) δ: 1.55-1.93 (6 H, m), 2.28-2.50 (4 H, m), 3.91 (2 H, m), 4.91 (2 H, d, J = 6); ^{13}C NMR (d^6-DMSO) δ: 26.4 (t), 30.8 (t), 37.0 (d), 70.7 (d).

9. Lautenschlaeger[5] has reported that diol **2** can be obtained in polymorphic forms with melting points between 188°C and 257°C according to the history of the sample. Determination of the mp is therefore not necessarily a good indicator of purity in this particular case, and the spectroscopic methods quoted are recommended. Crystals from methanol had mp 250-253°C (lit.[4] 249-250.5°C) but these rapidly became opaque and eventually crumbled to a white powder with lower mp. Samples of diol **2** prepared by the method described here melted indistinctly about 220-225°C.

10. Dimethyl sulfoxide was stirred and heated at ca. 100°C with powdered calcium hydride for 2 hr, distilled under reduced pressure, and used at once.

11. Care must be taken to ensure that the liquid from the funnel drops directly into the reaction and not onto the walls of the flask where it will solidify.

12. The physical properties of the product are as follows: mp 140-142°C; lit.[6] 140-142°C. IR (paraffin mull) cm^{-1}: 1690 (s), 1260 (m), 1215 (m), 1125 (m), 1095 (w), 1035 (w), 935 (w); ^{1}H NMR (CDCl$_3$) δ: 2.47-2.69 (6 H, m), 2.82-2.89 (2 H, m), 3.37-3.38 (2 H, m); ^{13}C NMR (CDCl$_3$) δ: 31.1 (t), 36.6 (t), 44.7 (d), 205.2 (s).

Waste Disposal Information

All toxic materials were disposed of in accordance with "Prudent Practices for Disposal of Chemicals from Laboratories"; National Academy Press; Washington, DC, 1983.

3. Discussion

The addition of sulfur dichloride and 1,5-cyclooctadiene to produce (1α,2α,5α,6α)-2,6-dichloro-9-thiabicyclo[3.3.1]nonane (**1**) has been described by several workers.[3-5,7] Determination of the stereochemistry of the product has been carried out using ^{1}H NMR methods,[3] and by means of the X-ray crystal structure of the corresponding sulfone derivative.[8] The procedure described here as step A is that due to Corey and Block.[3]

Hydrolysis of the dichloride (**1**) to yield (endo,endo)-9-thiabicyclo[3.3.1]nonane-2,6-diol (**2**) has been carried out using water,[4] aqueous sodium hydroxide,[4] aqueous sodium hydrogen carbonate,[5] and aqueous sodium carbonate.[5] Step B is an improved version of the latter reaction. The (endo,endo)-stereochemistry was originally inferred from IR evidence,[4] and subsequently confirmed by NMR work and the crystal structure of the related compound 2,6-dinitrato-9-thiabicyclo[3.3.1]nonane 9,9-dioxide.[9] Both the high lability of the dichloride (**1**) and the stereochemistry resulting from hydrolysis arise from neighboring group participation of the sulfur atom to form intramolecular sulfonium ion intermediates.[3,4] These processes have been studied in detail by Vincent and co-workers[10] using ^{13}C NMR spectroscopy.

The diketone (3) is a versatile intermediate for the preparation of 9-thiabicyclo[3.3.1]nonane derivatives[6,11-13] and provides a simple synthetic entry to a number of other heterocycles such as the 2-thiaadamantane,[13,14] thiacyclohexane,[15] thiacycloheptane,[15] 2,6-dithiaadamantane,[16] and 2-thiabrexane[17] ring systems. Only one previous procedure for its preparation has been published.[6] This method involves the oxidation of diol (2) with chromium trioxide in pyridine and dichloromethane to yield 9-thiabicyclo[3.3.1]nonane-2,6-dione (3) in 65% yield. In practice this reaction is difficult to carry out reproducibly because of precipitation of tarry chromium salts which make adequate stirring and extraction of the product very difficult. Consequently the dione (3) is often accompanied by partly oxidized material and/or material where the sulfur atom has been oxidized. The method reported here avoids these technical difficulties and affords a considerably increased yield.

More generally, the procedure described in step C illustrates how the Swern oxidation method[18-20] can be employed for the selective oxidation of an alcohol functionality in the presence of a sulfur moiety. A drawback of the original Swern oxidation is the lack of solubility of some substrates in the dichloromethane solvent at low temperatures, which results in a serious reduction of yield. In the past this has been avoided by carrying out the oxidation step at -10°C. This once again gave excellent yields, but this procedure required the use of twice the stoichiometric amount of oxidant.[19] The method described here as step C demonstrates that this is not necessary, and that the oxidation of insoluble materials can be carried out following the routine procedure provided that the substrate is added as a solution in dry dimethyl sulfoxide.

Step C has been carried out on three times the described scale without any deleterious effects being noted.

1. School of Chemistry, The University of New South Wales, Kensington, New South Wales 2033, Australia.
2. "Handbook of Preparative Inorganic Chemistry," 2nd English ed.; Brauer, G., Ed.; Academic Press: New York, 1963; Vol. 1, pp. 370-371.
3. Corey, E. J.; Block, E. *J. Org. Chem.* **1966**, *31*, 1663.
4. Weil, E. D.; Smith, K. J.; Gruber, R. J. *J. Org. Chem.* **1966**, *31*, 1669.
5. Lautenschlaeger, F. *Can. J. Chem.* **1966**, *44*, 2813.
6. MacNicol, D. D.; McCabe, P. H.; Raphael, R. A. *Synth. Commun.* **1972**, *2*, 185.
7. Dunlop Rubber Co. Ltd., French Patent No. 1 427 429, Feb. 4, 1966; *Chem. Abstr.* **1966**, *65*, P12181c.
8. McCabe, P. H.; Sim, G. A. *Acta Crystallogr., Sect. B* **1981**, *B37*, 1943.
9. McCabe, P. H.; Sim, G. A. *J. Chem. Soc., Perkin Trans. 2* **1982**, 819.
10. Vincent, J. A. J. M.; Schipper, P.; de Groot, Ae.; Buck, H. M. *Tetrahedron Lett.* **1975**, 1989.
11. Vincent, J. A. J. M.; Vereyken, F. P. M.; Wauben, B. G. M.; Buck, H. M. *Recl. Trav. Chim. Pays-Bas* **1976**, *95*, 236.
12. Hawkins, S. C.; Scudder, M. L.; Craig, D. C.; Rae, A. D.; Abdul Raof, R. B.; Bishop, R.; Dance, I. G. *J. Chem. Soc., Perkin Trans. 2* **1990**, 855.
13. McCabe, P. H.; Nelson, C. R.; Routledge, W. *Tetrahedron* **1977**, *33*, 1749.
14. McCabe, P. H.; Routledge, W. *Tetrahedron Lett.* **1973**, 3919.
15. McCabe, P. H.; Nelson, C. R. *Tetrahedron Lett.* **1978**, 2819.
16. Bishop, R. (manuscript in preparation).
17. Bishop, R.; Lee, G.-H.; Pich, K. C. (manuscript in preparation).
18. Omura, K.; Swern, D. *Tetrahedron* **1978**, *34*, 1651.
19. Mancuso, A. J.; Huang, S.-L.; Swern, D. *J. Org. Chem.* **1978**, *43*, 2480.
20. Mancuso, A. J.; Brownfain, D. S.; Swern, D. *J. Org. Chem.* **1979**, *44*, 4148.

Appendix

Chemical Abstracts Nomenclature (Collective Index Number); (Registry Number)

9-Thiabicyclo[3.3.1]nonane-2,6-dione (9); (37918-35-7)

(1α,2α,5α,6α)-2,6-Dichloro-9-thiabicyclo[3.3.1]nonane (9); (10502-30-4)

1,5-Cyclooctadiene: 1,5-Cyclooctadiene, (Z,Z)- (8,9); (1552-12-1)

Sulfur dichloride: Sulfur chloride (8,9); (10545-99-0)

(endo,endo)-9-Thiabicyclo[3.3.1]nonane-2,6-diol: 9-Thiabicyclo[3.3.1]nonane-2,6-diol stereoisomer (8,9); 9-Thiabicyclo[3.3.1]nonane-2,6-diol, (endo,endo)- (10); [22333-35-3]

Oxalyl chloride (8); Ethanedioyl dichloride (9); (79-37-8)

Dimethyl sulfoxide: Methyl sulfoxide (8); Methane, sulfinylbis- (9); (67-68-5)

4-BENZYL-10,19-DIETHYL-4,10,19-TRIAZA-1,7,13,16-TETRAOXACYCLOHENEICOSANE (TRIAZA-21-CROWN-7)

A. HO~~O~~NH$_2$ → (acetic anhydride / ethanol) → HO~~O~~NHCOCH$_3$

B. HO~~O~~NHCOCH$_3$ → (SOCl$_2$ / CHCl$_3$) → Cl~~O~~NHCOCH$_3$

C. Cl~~O~~NHCOCH$_3$ + C$_6$H$_5$CH$_2$NH$_2$ → 1) CH$_3$CN, NaI, Na$_2$CO$_3$; 2) BH$_3$-THF; 3) HCl, then NH$_4$OH → C$_6$H$_5$CH$_2$-N(CH$_2$CH$_2$OCH$_2$CH$_2$NHC$_2$H$_5$)$_2$

D. C$_6$H$_5$CH$_2$-N(CH$_2$CH$_2$OCH$_2$CH$_2$NHC$_2$H$_5$)$_2$ + I~~O~~O~~I → (K$_2$CO$_3$, CH$_3$CN) → triaza-21-crown-7 product

Submitted by Krzysztof E. Krakowiak[1] and Jerald S. Bradshaw.[2]
Checked by Hyunik Shin and James D. White.

1. Procedure

A. N-[2-(2-Hydroxyethoxy)ethyl]acetamide. Into an oven-dried, 100-mL, three-necked, round-bottomed flask that contains a magnetic stirring bar, 25-mL addition funnel, condenser, and thermometer, is placed 15.75 g (0.15 mol) of 2-(2-aminoethoxy)ethanol (Note 1) in 20 mL of anhydrous ethanol (Note 2). Acetic anhydride (15.75 g, 0.154 mol) (Note 1) is slowly dripped into this stirring solution while the temperature is kept at or below 40°C. The resulting mixture is stirred under reflux for 15 min. The mixture is evaporated on a rotary evaporator to give a pale yellow oil that is distilled through a short path apparatus containing a 2-cm Vigreux column to give 20.3-21.5 g (92-97%) of product, bp 135-139°C/0.12 mm (Note 3).

B. N-[2-(2-Chloroethoxy)ethyl]acetamide. N-[2-(2-Hydroxyethoxy)ethyl]acetamide (20.58 g, 0.14 mol) in 30 mL of chloroform (Note 4) is placed in an oven-dried, 250-mL, three-necked, round-bottomed flask that contains a magnetic stirring bar, pressure-equalizing dropping funnel, condenser that is connected by a polyethylene tube to a glass funnel, and a thermometer (Note 2). The flask is cooled in an ice bath and the glass funnel is immersed in a beaker of water. Thionyl chloride (28 g, 0.24 mol) (Note 1) in 25 mL of chloroform is added dropwise to the stirring mixture at 10-15°C. The mixture is then stirred at room temperature for 30 min and under reflux for 20 min (Note 5). It is immediately cooled to 30°C and the solvent and excess thionyl chloride are removed on a rotary evaporator (Note 6). The residue is immediately distilled through an 8-cm Vigreux column using an oil bath (Note 7) to give 17-18 g (73-77%) (Note 8) of N-[2-(2-chloroethoxy)ethyl]acetamide, bp 105-108°C/0.1 mm (Note 3).

C. 9-Benzyl-3,9,15-triaza-6,12-dioxaheptadecane. A mixture of 17.5 g (0.105 mol) of N-[2-(2-chloroethoxy)ethyl]acetamide, 200 mL of acetonitrile (Note 9), 22 g (0.21 mol) of anhydrous sodium carbonate (Note 10), 16.5 g (0.11 mol) of sodium iodide (Notes 1 and 11) and 5.35 g (0.05 mol) of benzylamine (Note 1) is added to an oven-dried, 250-mL, three-necked flask that contains a magnetic stirring bar and condenser. The mixture is refluxed for 48 hr, cooled to room temperature and filtered under vacuum. The solid in the filter is washed twice with 70-mL portions of methylene chloride (Note 12). The combined filtrate and methylene chloride mixture are evaporated on a rotary evaporator until all of the solvents are removed. The residue is dissolved in 70 mL of water and transferred to an extraction funnel. The aqueous solution is extracted with three 200-mL portions of methylene chloride (Note 12). The combined methylene chloride extracts are dried over anhydrous magnesium sulfate (Note 13). The mixture is filtered under vacuum and the solid is washed with 100 mL of methylene chloride. The solvents are evaporated on a rotary evaporator. The residue is dissolved in 80 mL of tetrahydrofuran (Note 14) and transferred to a dropping funnel equipped with an anhydrous calcium sulfate drying tube. This solution is slowly dripped into 300 mL of cold, stirring 1 N borane-THF (Notes 1 and 15) in an oven-dried, 1-L, one-necked, round-bottomed flask containing a magnetic stirring bar. The addition funnel is removed, a condenser equipped with an anhydrous calcium sulfate drying tube is connected, and the stirring mixture is refluxed for 16 hr. The mixture is cooled and 20 mL of distilled water is slowly dripped into it from a dropping funnel (Note 16). The solvents are then removed on a rotary evaporator. The residue is cooled in an ice water bath and 300 mL of aqueous 18% hydrochloric acid (Note 17) is slowly added to the stirring mixture. The resulting mixture is stirred for 16 hr, first at room temperature, and then warmed until the solvent just reaches the boiling point. The solution is cooled and the solvents are completely removed on a rotary evaporator. To the residue is added 50 mL of water. The aqueous mixture is stirred (Note 18) and filtered under vacuum. The solid is washed with 15 mL of water, and 150 mL of ammonium hydroxide (Note 19) is added to the filtrate (Note 2). The resulting solution is extracted three times with 300-mL portions of chloroform (35 g of sodium chloride is added to the aqueous solution before the third extraction). The chloroform layers are combined and dried over anhydrous magnesium sulfate. The solution is filtered under vacuum, the solid is washed with 50 mL of chloroform, and the solvent is removed on a rotary evaporator. The residue is distilled slowly through a short path apparatus that contains a 2-cm Vigreux column to give 11.6-12.6 g (69-75%) of 9-benzyl-3,9,15-triaza-6,12-dioxaheptadecane as a light yellow oil, bp 144°-152°/0.085 mm (Note 3).

D. *4-Benzyl-10,19-diethyl-4,10,19-triaza-1,7,13,16-tetraoxacycloheneicosane (triaza-21-crown-7)*. A mixture of 10 g (0.03 mol) of 9-benzyl-3,9,15-triaza-6,12-dioxaheptadecane and 600 mL of acetonitrile (Note 9) is placed in an oven-dried, 1-L, three-necked, round-bottomed flask equipped with a condenser, an efficient mechanical stirrer and a rubber septum. Argon gas is flushed through the flask using a needle through the septum before and during the reaction. To the above mixture, while stirring at room temperature, is added 50 g (0.36 mol) of anhydrous powdered potassium carbonate (Note 20) and the mixture is stirred for 15 min; then 11.5 g (0.031 mmol) of 1,2-bis(2-iodoethoxy)ethane (Note 1) is added. The resulting mixture is stirred under reflux for 24 hr. The cooled mixture is filtered under vacuum and the solid is washed with 100 mL of acetonitrile. The solvent is removed from the filtrate on a rotary evaporator. Methylene chloride (160 mL) (Note 12) is added to the residue and the resulting mixture is stirred (Note 18). The mixture is filtered under vacuum and the solid is washed with two 20-mL portions of methylene chloride (Note 12). The combined organic layers are evaporated on a rotary evaporator to give a brown oil. The oil is purified by chromatography on 300 g of alumina (Note 21) using 1000 mL of toluene/ethanol: 50:1 as the eluant. The first 120-150 mL of eluant is removed. The remainder of the eluant is evaporated in a rotary evaporator to give 9.6-10 g (72-75%) of the triaza-21-crown-7 as a light yellow oil (Notes 22-24).

2. Notes

1. 2-(2-Aminoethoxy)ethanol, acetic anhydride (99+%), thionyl chloride (99+%), benzylamine (99%), 1,2-bis(2-iodoethoxy)ethane (98%), anhydrous sodium iodide (99+%) and borane-tetrahydrofuran complex (1.0 M solution in tetrahydrofuran) were purchased from Aldrich Chemical Company, Inc. and were used without further purification.

2. The reaction must be carried out in an efficient hood.

3. The IR and NMR spectra in reference 3 were consistent with the proposed structure.

4. Chloroform ("Chrom Pure") manufactured by American Burdick & Jackson (distributed by American Scientific Products) was used.

5. The reaction also can be carried out at room temperature for 48 hr.

6. A rotary evaporator that has a safety flask in the vacuum line was used. The water bath temperature was not higher than 50°C.

7. The submitters prefer to use an oil bath rather than a heating mantle because control of pot temperature is essential so that no polymers can form. There is a loss of vacuum at the beginning of the distillation because of various gases in the product.

8. The yield from a seven times larger reaction was 85%.[3]

9. Acetonitrile ("Chrom Pure") manufactured by American Burdick & Jackson (distributed by American Scientific Products) was used.

10. Sodium carbonate (anhydrous A.C.S. certified) distributed by Fisher Scientific was used.

11. In reference 3, xylene was used as the solvent instead of acetonitrile and sodium iodide was not used. The yield in that case was lower.

12. Methylene chloride ("Chrom Pure") manufactured by American Burdick & Jackson (distributed by American Scientific Products) was used.

13. Magnesium sulfate (anhydrous, powder) distributed by EM Science was used.

14. Tetrahydrofuran (HPLC quality), manufactured by Mallinckrodt Inc., was used.

15. When fresh lithium aluminum hydride was used instead of the borane-tetrahydrofuran complex,[3] the reduction gave low yields and the product was more difficult to purify.

16. The first drops of water are added very slowly because the reaction causes considerable foaming.

17. Analytical reagent grade hydrochloric acid, manufactured by Mallinckrodt Inc., was used for the preparation of the 18% solution.

18. The mixture was stirred until the solid was completely suspended.

19. Analytical reagent grade ammonium hydroxide (NH_3 = 29.6%), manufactured by Mallinckrodt Inc., was used.

20. "Baker Analyzed" reagent grade potassium carbonate (anhydrous, granular) was used.

21. Activated, neutral, 150 mesh alumina (Brockman 1 standard grade), sold by Aldrich Chemical Company, Inc., was used in a 5-cm diameter column for the purification of the triaza-crown. Alumina TLC plates (Aluminum oxide 60_{F254} neutral (Type E), manufactured by E. Merck, distributed by EM Science) were used to monitor the purification process (toluene/ethanol = 20:1, using iodine as an indicator).

22. The product is nearly 98% pure as determined by GC using a Carlo Erba 5160 Mega instrument, flame ionization detection, with H_2 as the carrier gas (50 cm/sec). The column was a 10-m x 200-μm i.d. fused silica column coated at a 0.15-μm film thickness with SE-33 (methyl silicone). The split injector was kept at 300°C,

the detector at 280°C, and the column at 230°C. The product gave the correct elemental analysis and has the following spectral properties: ^{1}H NMR (CDCl$_3$, 200 MHz) δ: 1.0 (t, 6 H), 2.55 (m, 4 H), 2.72 (m, 12 H), 3.6 (m, 18 H), 7.35 (m, 5 H); IR (neat) cm^{-1}: 2860, 1440, 1340, 1110; MS (20 eV) m/e 160, 275, 336, 451 (100%).

23. The crown can be distilled under high vacuum in a short path distillation apparatus using an oil bath, bp 188-190°C/0.05 mm.

24. The product should be stored under argon at 0°C in a dark bottle to prevent decomposition.

Waste Disposal Information

All toxic materials were disposed of in accordance with "Prudent Practices for Disposal of Chemicals from Laboratories"; National Academy Press; Washington, DC, 1983.

3. Discussion

The procedure described here illustrates a new "building block" method for the preparation of per-N-alkyl-substituted polyaza-crown compounds.[3-5] The key building block, N-[2-(2-chloroethoxy)ethyl]acetamide and its benzamide analog,[3] allows the preparation of a variety of tri- and tetraamino compounds capable of ring closure reactions to form a variety of polyaza-crowns.[3,4] Polyaza-crowns containing a secondary amine side chain (mono lariat ether),[5] a hydroxyalkyl side chain,[6] and very large polyaza-crowns (30-36 ring members)[3] have been prepared using these procedures. The triaza-18-crown-6 analog was prepared in an overall yield of 25% by treating 9-benzyl-3,9,15-triaza-6,12-dioxaheptadecane with 3-oxapentanedioyl dichloride followed by reduction.[3,4]

The procedures previously used to prepare N-peralkylated polyaza-crowns required the use of nitrogen protecting groups that must subsequently be removed and the alkyl groups added.[7-11] These added steps greatly reduced the overall yields of the polyaza-crowns. Polyaza-crowns are important for complexing certain "soft" heavy metals,[12-16] and anions,[17-20] and as enzyme mimics in certain biological systems.[21] It is important to note that the N-peralkylated polyaza-crowns have about the same affinity for metal cations as the unsubstituted aza-crowns.[15] We have prepared 26 new N-peralkylated polyaza-crowns using this new method.

1. Department of Chemical Technology, School of Medicine, 90145 Lodz, Poland.
2. Department of Chemistry, Brigham Young University, Provo, UT 84602.
3. Krakowiak, K. E.; Bradshaw, J. S.; Izatt, R. M.; Zamecka-Krakowiak, D. J. *J. Org. Chem.* **1989**, *54*, 4061.
4. Krakowiak, K. E.; Bradshaw, J. S.; Izatt, R. M. *Tetrahedron Lett.* **1988**, *29*, 3521.
5. Bradshaw, J. S.; Krakowiak, K. E.; An, H.; Izatt, R. M. *Tetrahedron*, **1990**, *46*, 1163.
6. Bradshaw, J. S.; Krakowiak, K. E.; Izatt, R. M. *Tetrahedron Lett.* **1989**, *30*, 803.
7. Hosseini, M. W.; Lehn, J. M.; Duff, S. R.; Gu, K.; Mertes, M. P. *J. Org. Chem.* **1987**, *52*, 1662.
8. Graf, E.; Lehn, J.-M. *Helv. Chim. Acta* **1981**, *64*, 1040.
9. Dietrich, B.; Hosseini, M. W.; Lehn, J.-M.; Sessions, R. B. *Helv. Chim. Acta* **1983**, *66*, 1262.
10. Schmidtchen, F. P. *J. Org. Chem.* **1986**, *51*, 5161.
11. Krakowiak, K. E.; Bradshaw, J. S.; Zamecka-Krakowiak, D. J. *Chem. Rev.* **1989**, *89*, 929.
12. Izatt, R. M.; Bruening, R. L.; Bruening, M. L.; Tarbet, B. J.; Krakowiak, K. E.; Bradshaw, J. S.; Christensen, J. J. *Anal. Chem.* **1988**, *60*, 1825.

13. Bradshaw, J. S.; Krakowiak, K. E.; Tarbet, B. J.; Bruening, R. L.; Biernat, J. F.; Bochenska, M.; Izatt, R. M.; Christensen, J. J. *Pure Appl. Chem.*, **1989**, *61*, 1619.
14. Bradshaw, J. S.; Krakowiak, K. E.; Tarbet, B. J.; Bruening, R. L.; Griffin, L. D.; Cash, D. E.; Rasmussen, T. D.; Izatt, R. M. *Solvent Extr. Ion Exch.* **1989**, *7*, 855.
15. Izatt, R. M.; Bradshaw, J. S.; Nielsen, S. A.; Lamb, J. D.; Christensen, J. J.; Sen, D. *Chem. Rev.* **1985**, *85*, 271.
16. Comarmond, J.; Plumeré, P.; Lehn, J.-M.; Agnus, Y.; Louis, R.; Weiss, R.; Kahn, O.; Morgenstern-Badarau, I. *J. Am. Chem. Soc.* **1982**, *104*, 6330.
17. Dietrich, B.; Hosseini, M. W.; Lehn, J. M.; Sessions, R. B. *J. Am. Chem. Soc.* **1981**, *103*, 1282.
18. Hosseini, M. W.; Blacker, A. J.; Lehn, J.-M. *J. Chem. Soc., Chem. Commun.* **1988**, 596.
19. Hosseini, M. W.; Lehn, J.-M. *Helv. Chim. Acta* **1988**, *71*, 749.
20. Motekaitis, R. J.; Martell, A. E.; Murase, I. J.; Lehn, J.-M.; Hosseini, M. W. *Inorg. Chem.* **1988**, *27*, 3630.
21. Lehn, J.-M. *Science* **1985**, *227*, 849.

Appendix

Chemical Abstracts Nomenclature (Collective Index Number); (Registry Number)

N-[2-(2-Hydroxyethoxy)ethyl]acetamide: Acetamide, N-[2-(2-hydroxyethoxy)ethyl]- (12); (118974-46-2)

2-(2-Aminoethoxy)ethanol: Ethanol, 2-(2-aminoethoxy)- (8,9); (929-06-6)

Acetic anhydride (8); Acetic acid anhydride (9); (108-24-7)

N-[2-(2-Chloroethoxy)ethyl]acetamide: Acetamide, 2-(2-chloroethoxy)-N-ethyl- (9); (36961-73-6)

Thionyl chloride (8,9); (7719-09-7)

Benzylamine (8); Benzenemethanamine (9); (100-46-9)

Boran-Tetrahydrofuran: Furan, tetrahydro-, compd. with borane (1:1) (8,9); (14044-65-6)

1,2-Bis(2-iodoethoxy)ethane: Ethane, 1,2-bis(2-iodoethoxy)- (9); (36839-55-1)

TETRAHYDRO-3-BENZAZEPIN-2-ONES: LEAD TETRAACETATE OXIDATION OF ISOQUINOLINE ENAMIDES

Submitted by George R. Lenz and Ralph A. Lessor.[1]
Checked by Shaowo Liang and Leo A. Paquette.

1. Procedure

A. N-(tert-Butoxycarbonyl)-6,7-dimethoxy-1-methylene-1,2,3,4-tetrahydroisoquinoline. A 1-L, three-necked, round-bottomed flask is equipped with a thermometer, magnetic stirring bar, nitrogen inlet with gas bubbler, and a pressure-equalizing dropping funnel (Note 1). The flask is charged with 102.6 g (500 mmol) of 6,7-dimethoxy-1-methyl-3,4-dihydroisoquinoline (Note 2) and 150 mL of alcohol-free chloroform (Note 3). The mixture is warmed to 50°C with stirring. The dropping funnel is charged with a solution of 136.9 g (627 mmol) of di-tert-butyl pyrocarbonate (Note 4) in 50 mL of alcohol-free chloroform. The nitrogen is turned off, and the solution of di-tert-butyl pyrocarbonate is added to the mixture at such a rate as to maintain steady but controlled gas evolution (Note 5). The stirred reaction mixture is heated at 60-65°C until gas evolution ceases, then allowed to cool to room temperature with stirring.

Solvent is removed on a rotary evaporator at reduced pressure, and the resulting crude pink solid is dried overnight under high vacuum. The yield of slightly pink solid is 176.5-180 g (Notes 6, 7).

B. *N-(tert-Butoxycarbonyl)-7,8-dimethoxy-1,3,4,5-tetrahydro-2H-3-benzazepin-2-one.* A 1-L, three-necked, round-bottomed flask is equipped with a nitrogen inlet, mechanical stirrer, thermometer, and a pressure-equalizing dropping funnel (Note 8). The flask is charged with 93 g (210 mmol) of lead tetraacetate (Note 9) and 250 mL of glacial acetic acid. Stirring is started, the flask is immersed in an ice-water bath, and the funnel is charged with a solution of 61 g (200 mmol) of N-(tert-butoxycarbonyl)-6,7-dimethoxy-1-methylene-1,2,3,4-tetrahydroisoquinoline (Note 10) in 250 mL of methylene chloride. This solution is added at such a rate that a temperature of 19-23°C is maintained throughout the addition, typically over a period of 15 to 20 min. The cooling bath is removed, and the mixture is stirred at room temperature for 1 hr. Glycerol (4 mL) is added to quench unreacted lead tetraacetate, and the mixture is stirred for an additional 10 min.

The mixture is poured into 750 mL of water in a 2-L separatory funnel and shaken thoroughly. The phases are separated, and the aqueous phase is extracted with two 100-mL portions of methylene chloride. The combined organic layers are washed with 700 mL of water, followed by successive 100-mL portions of saturated aqueous sodium bicarbonate until no further effervescence is observed. The organic layer is dried over magnesium sulfate, filtered, and evaporated at reduced pressure to give a yellow-orange solid, which is dissolved in 100 mL of boiling acetone and allowed to cool slowly to room temperature, then kept overnight at -20°C. Filtration affords a cream-colored solid in a yield of 51.2-52.2 g (Notes 11, 12). A second crop is obtained by evaporation of the mother liquors, dissolution in a minimal amount of boiling acetone, cooling and seeding. The total yield is brought to 58-59 g (90-92%).

C. 7,8-Dimethoxy-1,3,4,5-tetrahydro-2H-3-benzazepin-2-one. A 250-mL, three-necked, round-bottomed flask is equipped with a magnetic stirring bar, thermometer, and pressure-equalizing dropping funnel. The flask is charged with 32.1 g (100 mmol) of recrystallized N-(tert-butoxycarbonyl)-7,8-dimethoxy-1,3,4,5-tetrahydro-2H-3-benzazepin-2-one and 100 mL of methylene chloride. The flask is immersed in an ice-water bath and stirred until the internal temperature reaches 5°C. The dropping funnel is charged with 40 mL of trifluoroacetic acid, which is added dropwise at a rate such that the temperature of the reaction mixture does not exceed 10°C (Note 13). When addition is complete, the cooling bath is removed, and the mixture is stirred for 3 hr (Note 14).

The mixture is diluted with an additional 700 mL of methylene chloride and poured into a separatory funnel containing 500 mL of water. The funnel is shaken thoroughly, and the phases are separated. The organic layer is washed with 500 mL of water, followed by successive washes with 200-mL portions of saturated aqueous sodium bicarbonate until no further effervescence is observed (Note 15). The organic phase is dried over sodium sulfate and evaporated under reduced pressure to give 24.8-25.2 g of crude product. The crude material is dissolved in a minimal amount of boiling methylene chloride, and then diluted with an equal volume of ethyl acetate. The mixture is boiled down to approximately two thirds of its starting volume, by which time crystallization has begun, then allowed to cool slowly to room temperature. After storage at -20°C overnight, filtration affords 17.5-18.0 g (79-81%) of product, mp 194-195°C (Notes 16, 17).

2. Notes

1. The glassware is dried in an oven at 110°C and assembled while still hot, then allowed to cool while a slow stream of nitrogen is passed through the apparatus.

2. 6,7-Dimethoxy-1-methyl-3,4-dihydroisoquinoline was prepared according to an *Organic Syntheses* procedure: Brossi, A.; Dolan, L. A.; Teitel, S. *Org. Synth., Coll. Vol. VI* **1988**, 1.

3. The submitters used J. T. Baker Chemical Company hydrocarbon-stabilized chloroform containing 0.015% amylene stabilizer. The checkers used chloroform of comparable quality purchased from Aldrich Chemical Company, Inc.

4. The submitters used commercially available tert-butyl pyrocarbonate from either Fluka or Aldrich Chemical Company, Inc. Alternatively, the reagent can be prepared according to Pope, B. M.; Yamamoto, Y.; Tarbell, D. S. *Org. Synth., Coll. Vol. VI* **1988**, 418.

5. The addition typically took 1.5 to 2 hr. The bubbler on the nitrogen line used to flush the flask is conveniently used to monitor the evolution of carbon dioxide as the reaction proceeds.

6. This material is pure enough for use in the next step. The impurities of tert-butyl alcohol and a small amount of unreacted tert-butyl pyrocarbonate do not interfere with the oxidation.

7. If desired, the material can be recrystallized from methanol; under these circumstances, 146-149 g of white solid, mp 101-102°C, is returned. The spectral characteristics of recrystallized material are as follows: ^{1}H NMR (300 MHz, CDCl$_3$) δ: 1.49 (s, 9 H), 2.78 (t, 2 H, J = 5.9), 3.77 (t, 2 H, J = 5.9), 3.86 (s, 3 H), 3.89 (s, 3 H), 5.31 (s, 1 H), 5.50 (s, 1 H), 6.56 (s, 1 H), 7.11 (s, 1 H); IR (KBr) cm^{-1}: 1690, 1630, 1605, 1510, 1390, and 1170; ^{13}C NMR (75 MHz, CDCl$_3$) δ: 28.38, 28.96, 43.51, 55.89, 56.02, 80.47, 101.86, 107.46, 110.77, 124.84, 127.72, 139.99, 147.51, 149.22, 153.83.

Anal. Calcd. for $C_{17}H_{23}NO_4$: C, 66.86; H, 7.59; N, 4.59. Found: C, 67.02; H, 7.48; N, 4.65.

8. The glassware is assembled hot under nitrogen as for the previous step. The nitrogen inlet and stirrer are mounted on a Claisen adapter, and the thermometer is removed during the flushing period, then reinserted.

9. The submitters used lead tetraacetate from Aldrich Chemical Company, Inc., which was dried under reduced pressure at room temperature for 10 min prior to use to remove any acetic acid present. Alternatively, lead tetraacetate still containing acetic acid may be used successfully if a slight excess is used.

10. If crude material containing tert-butyl alcohol and unreacted pyrocarbonate is used in this step, the amount of starting material present is calculated based on the mass balance for the first step, assuming a quantitative conversion, and 1.05 equivalents of lead tetraacetate are used. The checkers used only pure material and advise against carrying forward less pure carbamate.

11. This material may be used directly in the following step. If desired, the material can be recrystallized from acetone, mp 116.5-118°C. The spectral characteristics of the recrystallized material are as follows: ^{1}H NMR (300 MHz, $CDCl_3$) δ: 1.52 (s, 9 H), 3.14 (t, 2 H, J = 6.0), 3.84 (s, 6 H), 3.92 (s, 2 H), 4.18 (t, 2 H, J = 6.0), 6.56 (s, 1 H), 6.57 (s, 1 H); IR (KBr) cm^{-1}: 1715, 1610, 1525, 1370, 1255, 1110, and 1060; ^{13}C NMR (75 MHz, $CDCl_3$) δ: 28.03, 32.83, 43.42, 45.21, 55.94, 55.97, 83.18, 113.25, 114.28, 121.92, 127.09, 147.41, 148.38, 152.08, 171.33. Anal. Calcd. for $C_{17}H_{23}NO_5$: C, 63.54; H, 7.21; N, 4.36. Found: C, 63.36; H, 7.30; N, 4.16.

12. If insufficient lead tetraacetate is used in the oxidation, unoxidized starting enamide is hydrolyzed during the workup to tert-butyl 2-(2-acetyl-3,4-dimethoxyphenyl) ethyl carbamate, mp 111.5-112.5°C (cf. Note 2): ^{1}H NMR (270 MHz,

$$\text{CH}_3\text{O}\underset{\text{CH}_3\text{O}}{\diagdown}\underset{\underset{\text{O}}{\overset{\|}{\text{C}}}\text{CH}_3}{\bigcirc}\diagup\text{NHCO}_2\text{C}(\text{CH}_3)_3$$

CDCl$_3$) δ: 1.42 (s, 9 H), 2.58 (s, 3 H), 3.03 (t, 2 H), 3.37 (q, 2 H), 3.92 (s, 3 H), 6.76 (s, 1 H), 7.23 (s, 1 H); IR (FTIR) cm^{-1}: 1707, 1674, 1604, 1517, 1266, 1212, 1152. Anal. Calcd for C$_{17}$H$_{25}$NO$_5$: C, 63.13; H, 7.79; N, 4.33. Found: C, 63.24; H, 7.91; N, 4.28. The hydrolyzed material co-migrates with the oxidation product in a variety of TLC systems, and also co-crystallizes with it. It is, however, removed during the trifluoroacetic acid (TFA) cleavage to form the benzazepinone (Step C). The presence of any hydrolyzed material is readily detected by the presence of the acetyl resonance (δ 2.58) in the NMR spectrum.

13. After approximately 25 mL of the trifluoroacetic acid have been added, gas evolution begins. This can be quite vigorous if the temperature is not kept below 10°C.

14. The progress of the reaction can be monitored by thin layer chromatography on silica gel plates, using a 95:5:0.5 mixture of chloroform:methanol:concentrated ammonium hydroxide as the developing solvent.

15. Foaming can be quite vigorous, especially if the reaction mixture is not washed first with water prior to the use of sodium bicarbonate solution.

16. A small second crop of impure material can be obtained from the mother liquors.

17. The product exhibited the following spectral characteristics: ^{1}H NMR (300 MHz, CDCl$_3$) δ: 3.03 (t, 2 H, J = 6.0), 3.52-3.60 (m, 2 H), 3.75 (s, 2 H), 3.83 (s, 3 H), 3.84 (s, 3 H), 6.34 (br s, 1 H), 6.59 (s, 1 H), 6.62 (s, 1 H); ^{13}C NMR (75 MHz, CDCl$_3$) δ: 33.03, 40.62, 41.71, 55.77 (2C), 112.96, 113.53, 123.19, 128.39, 147.09, 147.72, 174.49; IR (KBr) cm^{-1}: 1675, 1220, 1125, 1100, and 1010.

Waste Disposal Information

All toxic materials were disposed of in accordance with "Prudent Practices for Disposal of Chemicals from Laboratories"; National Academy Press; Washington, DC, 1983.

3. Discussion

This procedure illustrates a general route to tetrahydro-3-benzazepin-2-ones from readily available dihydroisoquinolines.[2] The benzazepine ring system exists in various classes of isoquinoline-derived alkaloids,[3] while other members of this class are being developed as pharmaceutical agents.[4] The present procedure takes advantage of ready formation of enamides from dihydroisoquinolines and carboxylic acid anhydrides, acid chlorides, carbonic anhydrides and chlorides and their facile oxidation to differentially functionalized benzazepinones (Table 1). The mechanism has been described and involves migration of the isoquinoline aromatic ring to the exocyclic methylene group.[2]

Several approaches to the synthesis of the tetrahydrobenzazepine ring system have been described,[5] and excellent methods exist for the preparation of aryl substituted tetrahydrobenzazepines.[4] However, benzazepines that are either unsubstituted or alkyl-substituted on the azepine ring are much less readily obtainable. For instance, the benzazepinone, synthesized by this procedure, was originally isolated, in low yield, from the mixture of photoproducts obtained from the irradiation of N-[3-(3,4-dimethoxyphenyl)]propyl chloroacetamide.[6] Preparative approaches to the benzazepinones have required multiple steps starting from an N-phenylethylacetamide and involving chloromethylation, cyanide displacement, nitrile solvolysis, hydrolysis to the amino acid and cyclization.[7] The 1-alkyl derivatives are subsequently prepared by alkylation of the parent compound.[8] The current procedure reduces the preparation of the tetrahydrobenzazepinone ring system to two straightforward steps.

Ring expansion of the isoquinoline enamides is insensitive to the type of acyl functionality used to form the enamide.[9] The reaction occurs when the isoquinoline aromatic ring is unsubstituted, or contains electron releasing substituents. The reaction is sensitive, however, to the degree and type of substitution on the exocyclic methylene group. Oxidative ring expansion occurs when the double bond is either unsubstituted or monoalkyl substituted. Phenyl substitution yields differing products depending on a number of variables.[9] When the exocyclic double bond is disubstituted, oxidation with lead tetraacetate proceeds readily, but does not lead to ring expansion.[10] The ring expansion reaction works equally well for the preparation of tetrahydrobenzazocinones from tetrahydrobenzazepine enamides (Table 2).[11]

The acid-catalyzed cleavage of the tert-butoxycarbonyl group is the best method to form the parent benzazepinone. Other methods used have been the Pd/C hydrogenolysis of a benzyloxycarbonyl group,[2,11,12] and zinc mediated reductive cleavage of trichloroethoxy and trichloro-tert-butoxy carbonyl groups.[11]

1. Health Care Research and Development, The BOC Group Technical Center, 100 Mountain Avenue, Murray Hill, NJ 07974.
2. Lenz, G. R. *J. Org. Chem.* **1988**, *53*, 5791-5793.
3. (a) Rönsch, H. In "The Alkaloids: Chemistry and Pharmacology"; Brossi, A., Ed.; Academic Press: New York, 1986; Vol. 28, pp. 1-93; (b) Montgomery, C. T.; Cassels, B. K.; Shamma, M.; *J. Nat. Prod.* **1983**, *46*, 441-453; (c) Fajardo, V.; Elango, V.; Cassels, B. K.; Shamma, M. *Tetrahedron Lett.* **1982**, *23*, 39-42.
4. Weinstock, J.; Hieble, J. P.; Wilson, J. *Drugs Fut.* **1985**, *10*, 646-697.
5. Kasparek, S. *Adv. Heterocycl. Chem.* **1974**, *17*, 45-98.

6. (a) Yonemitsu, U.; Okuno, Y.; Kanaoka, Y.; Karle, I. L.; Witkop, B. *J. Am. Chem. Soc.* **1968**, *90*, 6522-6523; (b) Yonemitsu, O.; Okuno, Y.; Kanaoka, Y.; Witkop, B. *J. Am. Chem. Soc.* **1970**, *92*, 5686-5690.

7. (a) Pecherer, B.; Sunbury, R. C.; Brossi, A. *J. Heterocycl. Chem.* **1972**, *9*, 609-616, 617-621; (b) Orito, K.; Kaga, H.; Itoh, M.; De Silva, S.; Manske, R. H.; Rodrigo, R. *J. Heterocycl. Chem.* **1980**, *17*, 417-423.

8. Orito, K.; Matsuzaki, T. *Tetrahedron* **1980**, *36*, 1017-1021.

9. Lenz, G. R.; Costanza, C. *J. Org. Chem.* **1988**, *53*, 1176-1183.

10. Lenz, G. R., *J. Chem. Soc., Perkin Trans. 1* **1990**, 33-38.

11. Lessor, R. A.; Rafalko, P. W.; Lenz, G. R. *J. Chem. Soc., Perkin Trans. 1* **1989**, 1931-1938.

12. Lenz, G. R. *Heterocycles* **1987**, *26*, 721-730.

Table 1

Lead Tetraacetate Oxidative Ring Expansion of Isoquinoline Enamides

Substrate	Product	Yield (%)
		62
		91
		92
		94
		50
		72

Table 2

Lead Tetraacetate Oxidative Ring Expansion of Benzazepine Enamides

Starting Material	Product	Yield (%)
6,7-dimethoxy-1-methylene-2-acetyl-benzazepine	ring-expanded N-COCH$_3$ ketone	76
6,7-dimethoxy-1-methylene-2-benzoyl-benzazepine	ring-expanded N-COC$_6$H$_5$ ketone	83
6,7-dimethoxy-1-methylene-2-(t-butoxycarbonyl)-benzazepine	ring-expanded N-CO$_2$C(CH$_3$)$_3$ ketone	73
6,7-dimethoxy-1-methylene-2-(benzyloxycarbonyl)-benzazepine	ring-expanded N-CO$_2$CH$_2$C$_6$H$_5$ ketone	78

Appendix

Chemical Abstracts Nomenclature (Collective Index Number); (Registry Number)

N-(tert-Butoxycarbonyl)-6,7-dimethoxy-1-methylene-1,2,3,4-tetrahydroisoquinoline: 2(1H)-Isoquinolinecarboxylic acid, 3,4-dihydro-6,7-dimethoxy-1-methylene, 1,1-dimethylethyl ester (11); (82044-08-4)

6,7-Dimethoxy-1-methyl-3,4-dihydroisoquinoline: Isoquinoline, 3,4-dihydro-6,7-dimethoxy-1-methyl- (8,9); (4721-98-6)

Di-tert-butyl pyrocarbonate: Formic acid, oxydi-, di-tert-butyl ester (8); Dicarbonic acid, bis(1,1-dimethylethyl) ester (9); (24424-99-5)

Lead tetraacetate: Acetic acid, lead (4+) salt (8,9); (546-67-8)

Glycerol (8); 1,2,3-Propanetriol (9); (56-81-5)

7,8-Dimethoxy-1,3,4,5-tetrahydro-2H-3-benzazepin-2-one: 2H-3-Benzazepin-2-one, 1,3,4,5-tetrahydro-7,8-dimethoxy- (8,9); (20925-64-8)

Trifluoroacetic acid: Acetic acid, trifluoro- (8,9); (76-05-1)

2-SUBSTITUTED PYRROLES FROM N-tert-BUTOXYCARBONYL-2-BROMOPYRROLE: N-tert-BUTOXY-2-TRIMETHYLSILYLPYRROLE

Submitted by Wha Chen, E. Kyle Stephenson, Michael P. Cava,[1] and Yvette A. Jackson.[2]
Checked by Wei He and Leo Paquette.

1. Procedure

A. N-tert-Butoxycarbonyl-2-bromopyrrole. A dry, 500-mL, three-necked, round-bottomed flask is equipped with a magnetic stirring bar, two solid addition funnels, and a three-way stopcock attached to a balloon filled with nitrogen. To the flask are added 4.5 g (67.2 mmol) of pyrrole (Note 1) and 180 mL of tetrahydrofuran (Note 2). The flask is evacuated and purged with nitrogen (Note 3). The stirred solution is cooled to -78°C with a dry ice-acetone bath (Note 4) and a catalytic amount (ca. 0.1 g) of azoisobutyronitrile (AIBN) (Note 5) is added via solid addition funnel. After 5 min, 9.57 g (33.6 mmol) of 1,3-dibromo-5,5-dimethylhydantoin (Note 6) is added over a 20-min period via solid addition funnel. The light-green mixture is stirred for an additional 10 min, then allowed to stand for 2 hr, keeping the temperature below -50°C. The solution is filtered by suction into a dry, 500-mL, round-bottomed flask that has been cooled to -78°C in a dry ice-acetone bath. The flask is equipped with a magnetic stirring bar and a three-way stopcock attached to a balloon filled with nitrogen. To the stirred dark-green solution is added 2.71 g (26.9 mmol) of triethylamine followed immediately by addition of 20.4 g (93.9 mmol) of di-tert-butyl dicarbonate and a catalytic amount (ca. 0.1 g) of 4-dimethylaminopyridine (Note 7). The flask is evacuated and purged with nitrogen (Note 3). The mixture is stirred for 8 hr while it is allowed to warm to room temperature (Note 8). The solvent is removed under reduced pressure at room temperature and 100 mL of hexane is added to the crude product, which is washed with deionized water (3 x 100 mL), dried over sodium sulfate, and concentrated under reduced pressure at room temperature. The crude product is purified by chromatography on amine-treated neutral silica (270 g) using hexane as the eluent (Note 9). The fractions containing the product are identified by TLC, combined, and concentrated under reduced pressure at room temperature to yield compound **1** as a colorless oil (13.5-14.7 g, 82-89%) (Note 10).

B. *N-tert-Butoxycarbonyl-2-trimethylsilylpyrrole.* A solution of N-tert-butoxycarbonyl-2-bromopyrrole (13.5 g, 54.9 mmol) in 40 mL of hexane is added to 200 mL of tetrahydrofuran (Note 2) in a dry, 500-mL, two-necked, round-bottomed flask equipped with a magnetic stirring bar, rubber septum, and a three-way stopcock attached to a balloon of nitrogen. The flask is evacuated and purged with nitrogen (Note 3). The stirred mixture is cooled to -78°C and 34.3 mL of 1.6 M butyllithium in hexane (Note 10) is added slowly via syringe over a 10-min period, during which time the colorless solution becomes brown. After an additional 10 min, 13.4 g (124 mmol) of chlorotrimethylsilane (Note 11) in 10 mL of tetrahydrofuran (Note 3) is added via syringe over a 10-min period. Stirring is continued and the mixture is allowed to warm to -30°C over a 1-hr period. The reaction mixture is quenched with saturated aqueous sodium bicarbonate (10 mL) at which point a dark red-purple color develops. After warming to 0°C, the solvent is removed under reduced pressure and the product is extracted into 300 mL of hexane. The organic layer is washed twice with 150 mL of water and then dried over anhydrous sodium sulfate. The solvent is removed under reduced pressure, and the residue is distilled twice using a Kugelrohr oven at 85°C and 0.15 mm to give the pure product **2** (10.5-11.1 g, 80-85%) (Note 12).

2. Notes

1. Pyrrole (Aldrich Chemical Company, Inc.) was freshly distilled before use.

2. Tetrahydrofuran was distilled from sodium benzophenone ketyl.

3. The apparatus is maintained under a nitrogen atmosphere during the course of the reaction.

4. The level of the reaction mixture must remain below the level of the cooling bath to avoid partial decomposition of the bromination product.

5. Azoisobutyronitrile (AIBN) (Fluka) was used as received.

6. Commercial 1,3-dibromo-5,5-dimethylhydantoin (Aldrich Chemical Company, Inc.) (22.0 g) was stirred for 12 hr at room temperature with 400 mL of 5% aqueous sodium bicarbonate, then stirred with 400 mL of deionized water for 8 hr, filtered, washed with 500 mL of deionized water and dried over phosphorus pentoxide to constant weight. The checkers used the commercial brominating agent as received from Aldrich Chemical Company, Inc.

7. Di-tert-butyl dicarbonate (Aldrich Chemical Company, Inc.) and 4-dimethylaminopyridine (Aldrich Chemical Company, Inc.) were used as received.

8. The checkers found that the reaction mixture must be stirred at room temperature for at least 2 hr prior to workup. It is advisable to monitor the progress of reaction by TLC.

9. The column is packed with hexane and pretreated with 500 mL of 5% triethylamine in hexane, then washed with 700 mL of hexane before addition of the compound.

10. Although this N-BOC derivative is far more stable than 2-bromopyrrole, it is best stored as a 20-25% solution in hexane at -10°C. Under these conditions, solutions show no sign of decomposition after many months. The product shows the following spectrum: ^{1}H NMR (CDCl$_3$) δ: 1.61 (s, 9 H), 6.14 (t, 1 H, J = 3.5), 6.29 (dd, 1 H, J = 2.0, 3.5), 7.30 (dd, 1 H, J = 2.0, 3.5).

11. Butyllithium solution (Aldrich Chemical Company, Inc.) and chlorotrimethylsilane (Aldrich Chemical Company, Inc.) were used as received.

12. The spectral properties for 2-trimethylsilyl-N-BOC pyrrole are as follows: ^{1}H NMR (CDCl$_3$) δ: 0.27 (s, 9 H), 1.60 (2, 9 H), 6.21 (t, 1 H, J = 3.0), 6.46 (dd, 1 H, J = 1.5, 3.0), 7.38 (dd, 1 H, J = 1.5, 3.0).

Waste Disposal Information

All toxic materials were disposed of in accordance with "Prudent Practices for Disposal of Chemicals from Laboratories"; National Academy Press; Washington, DC, 1983.

3. Discussion

Whereas 2-lithiothiophene and 2-lithiofuran are readily prepared from butyllithium and the parent heterocycles by lithium-hydrogen exchange, a similar exchange with pyrrole affords only N-lithiopyrrole. A study of the lithium-hydrogen exchange of several N-blocked pyrroles with strong bases concluded that synthetically useful lithium-hydrogen exchange at the 2-position could best be effected using as the substrate N-tert-butoxycarbonylpyrrole, but only in conjunction with the very hindered and costly base lithium tetramethylpiperidide.[3]

In the present procedure, pyrrole is brominated under mild conditions to the very labile 2-bromopyrrole using 1,3-dibromo-5,5-dimethylhydantoin: the latter reagent gives better results than the previously employed N-bromosuccinimide.[4] Direct conversion of 2-bromopyrrole to its more stable N-tert-butoxycarbonyl derivative (1) affords a substrate which readily undergoes lithium-halogen exchange with butyllithium at -78°C. Subsequent reaction with an electrophile is exemplified by the reaction with chlorotrimethylsilane to give N-tert-butoxycarbonyl-2-trimethylsilylpyrrole (2). Other electrophiles (e.g., dimethyl disulfide, methyl chloroformate) have also been employed successfully.[5] In addition, a similar procedure has been used to convert pyrrole into N-tert-butoxy-2,5-disubstituted pyrroles.[5]

The N-tert-butoxycarbonyl protecting group of substituted pyrroles can be removed readily by methoxide ion[3] or, when electron-withdrawing substituents are present, by mild thermolysis.[6]

1. Department of Chemistry, The University of Alabama, P.O. Box 870336, Tuscaloosa, AL 35487-0336.
2. Department of Chemistry, University of the West Indies, Mona, Kingston 7, Jamaica, West Indies.
3. Hasan, I.; Marinelli, E. R.; Chang Lin, L.-C.; Fowler, F. W.; Levy, A. B. *J. Org. Chem.* **1981**, *46*, 157.
4. Gilow, H. W.; Burton, D. E. *J. Org. Chem.* **1981**, *46*, 2221.
5. Chen, W.; Cava, M. P. *Tetrahedron Lett.* **1987**, *28*, 6025.
6. Rawal, V. H.; Cava, M. P. *Tetrahedron Lett.* **1985**, *26*, 6141.

Appendix

Chemical Abstracts Nomenclature (Collective Index Number); (Registry Number)

N-tert-Butoxycarbonyl-2-bromopyrrole: 1H-Pyrrole-1-carboxylic acid, 2-bromo-, 1,1-dimethylethyl ester (12); (117657-37-1)

N-tert-Butoxycarbonyl-2-trimethylsilylpyrrole: 1H-Pyrrole-1-carboxylic acid, 2-(trimethylsilyl)-, 1,1-dimethylethyl ester (10); (75400-57-6)

Pyrrole: 1H-Pyrrole (9); (109-97-7)

Azobisisobutyronitrile: Propionitrile, 2,2'-azobis[2-methyl- (8); Propanenitrile, 2,2'-azobis[2-methyl- (9); (78-67-1)

1,3-Dibromo-5,5-dimethylhydantoin: Hydantoin, 1,3-dibromo-5,5-dimethyl- (8); 2,4-Imidazolidinedione, 1,3-dibromo-5,5-dimethyl- (9); (77-48-5)

Di-tert-butyl dicarbonate: Formic acid, oxydi-, di-tert-butyl ester (8); Dicarbonic acid, bis(1,1-dimethylethyl) ester (9); (24424-99-5)

Chlorotrimethylsilane: Silane, chlorotrimethyl- (8,9); (75-77-4)

SUBSTITUTION REACTIONS OF 2-BENZENESULFONYL CYCLIC ETHERS: TETRAHYDRO-2-(PHENYLETHYNYL)-2H-PYRAN

(2H-Pyran, tetrahydro-2-(phenylethenyl)-)

Submitted by Dearg S. Brown and Steven V. Ley.[1]
Checked and modified by David J. Mathre and Ichiro Shinkai.

1. Procedure

An oven-dried, three-necked, 1-L, round-bottomed Morton flask equipped with a mechanical stirrer, gas bubbler outlet, 125-mL, pressure-equalizing addition funnel fitted with a rubber septum, and a nitrogen inlet (Note 1) is charged with 11.2 g (110 mmol) of phenylacetylene (Note 2) and 60 mL of dry tetrahydrofuran (THF) (Note 3). The addition funnel is charged with 60 mL of 2 M isopropylmagnesium chloride (Note 4) which is then added over a 5-min period (Note 5). The addition funnel is rinsed with 10 mL of dry THF and the solution is stirred at room temperature for 1 hr. The addition funnel is charged with 72 mL of 1 M anhydrous zinc bromide in THF (Note 6) which is then added to the light-grey solution over a 5-min period (Note 7). The addition funnel is rinsed with 10 mL of dry THF, and the mixture is stirred for a further 30 min at room temperature (Note 8). The addition funnel is charged with a solution of 22.6 g (100 mmol) of 2-(phenylsulfonyl)tetrahydro-2H-pyran (Note 2) dissolved in 100 mL of dry THF, which is then added over a 5-min period (Note 9). The resulting grey solution is stirred at room temperature for 18 hr and then quenched with 300 mL of 1 M hydrochloric acid (Note 10). The mixture is transferred to a single necked, 1-L, round-bottomed flask and concentrated under reduced pressure (40°C, 100 mm) to remove the THF. The residue is transferred to a 1-L separatory funnel and extracted with 300 mL of isopropyl acetate (i-PrOAc). The extract is sequentially washed with water (150 mL), 1 M aqueous dibasic potassium phosphate (3 x 150 mL), and brine (150 mL). The extract is dried over anhydrous sodium sulfate, filtered through a sintered glass funnel, washing the residue with more i-PrOAc, and then concentrated under reduced pressure (40°C, 10 mm) to give 19.6 g of crude product as a pale yellow liquid (Note 11). This is distilled under reduced pressure to afford 18.3 g (98%) of tetrahydro-2-(phenylethynyl)-2H-pyran as a colorless liquid, bp 110-115°C (0.01 mm) (Note 12).

2. Notes

1. A constant stream of anhydrous nitrogen (dried over molecular sieves) was maintained throughout the reaction.

2. All the chemicals used in this procedure were purchased from Aldrich Chemical Company, Inc., and were used without further purification unless otherwise stated.

3. Tetrahydrofuran was dried over 3Å molecular sieves (residual water content <20 µg/mL) and purged with nitrogen prior to use.

4. Isopropylmagnesium bromide can also be used in this type of experiment.

5. The reaction is exothermic, the internal temperature rising from 20°C to 44°C. *Caution: Do not run the reaction under more concentrated conditions, or on a larger scale without providing external cooling.*

6. Zinc bromide (100 g, 440 mmol) (98+%, Note 2) was dissolved in dry THF bringing the volume to 440 mL (exothermic heat of solution). The initial water content

(1.2 mg/mL) was reduced to <50 μg/mL by drying the solution with 3Å molecular sieves (50 g) for 24 hr.

7. The reaction is exothermic, the internal temperature rising from 20°C to 34°C. See Note 5.

8. On addition of zinc bromide a fine white precipitate is sometimes formed (observed by checkers).

9. The reaction is exothermic, raising the internal temperature from 20°C to 32°C. See Note 5.

10. The quench is exothermic during addition of the first ca. 25 mL of the aqueous hydrochloric acid, the internal temperature rising from 20°C to 30°C.

11. HPLC analysis is as follows: 94.1 wt% product, 2.4 wt% phenylacetylene, remainder i-PrOAc. HPLC conditions are as follows: [4.6 x 250 mm Zorbax RX; 40:60 H_2O (0.01 M KH_2PO_4)/MeCN; 1.5 mL/min; UV 210 nm] phenylsulfinic acid (2.6 min), 2-(phenylsulfonyl)tetrahydro-2H-pyran (4.7 min, decomposes in solution), phenylacetylene (6.1 min), tetrahydro-2-(phenylethynyl)-2H-pyran (9.2 min).

12. The physical properties are as follows: literature bp 149°C (8 mm);[2] HPLC analysis: >99 wt% product, <0.1% phenylacetylene; ^1H NMR ($CDCl_3$) δ: 1.50-2.00 (m, 6 H, C3-H_2, C4-H_2, C5-H_2), 3.53-3.66 (m, 1 H, C6-H), 4.00-4.13 (m, 1 H, C6-H), 4.52 (dd, 1 H, J = 2.8, 7.4, C2-H), 7.26-7.35 (m, 3 H, Ar-H), 7.41-7.51 (m, 2 H, Ar-H); ^{13}C NMR ($CDCl_3$) δ: 21.8 (t), 25.7 (t), 32.2 (t), 66.6 (t), 67.5 (d), 85.2 (s), 88.1 (s), 122.8 (s), 128.2 (d), 128.3 (d), 131.8 (d).

Waste Disposal Information

All toxic materials were disposed of in accordance with "Prudent Practices for Disposal of Chemicals from Laboratories"; National Academy Press; Washington, DC, 1983.

3. Discussion

Methods for forming carbon-carbon bonds at the anomeric position of cyclic ethers are important processes in organic synthesis. We have shown how lactols and their derivatives can be readily converted into the corresponding 2-benzenesulfonyl cyclic ethers.[3,4] These versatile intermediates can then be transformed into the corresponding dihydropyrans,[3] 2-substituted dihydropyrans,[4] spiroacetals,[4,5] and tetrahydropyranyl ethers[6] (Scheme 1).

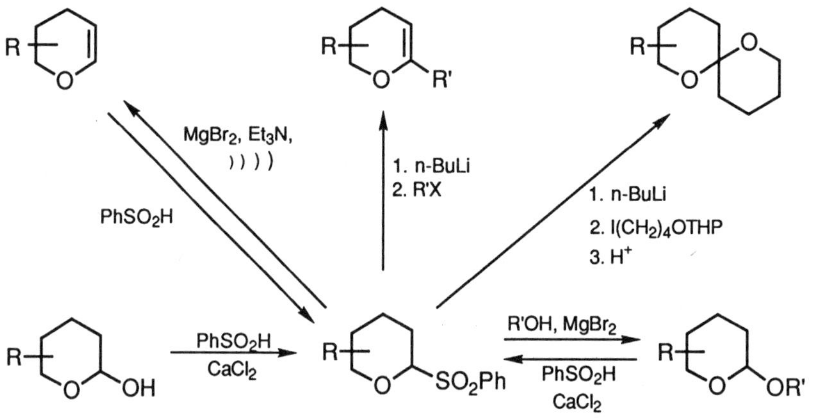

Scheme 1

The procedure illustrated here is representative of a general and versatile method for the preparation of 2-substituted tetrahydrofurans and tetrahydropyrans from cyclic ether sulfones and the appropriate alkynyl, vinyl, or aryl Grignard reagent. From the examples shown in the Table and others previously reported,[3,7] a selectivity for the trans-product is observed with 6-substituted tetrahydropyrans irrespective of the initial geometry of the sulfone. This implies the presence of a common reaction intermediate such as an oxonium ion which is trapped by preferred axial bond formation at the 2-position. For 5-substitued tetrahydrofurans, significant trans-selectivity is only observed with large substituents.

TABLE

SUBSTITUTION REACTIONS OF 2-BENZENESULFONYL CYCLIC ETHERS

Sulfone	Product(s)	Yield (%)
THP-SO₂Ph	THP-C≡C-CH₂CH₂-OTHP	83
Ph-THP-SO₂Ph	Ph-THP-C≡C-n-Bu	81
MeO-THF-SO₂Ph	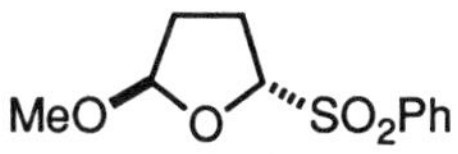 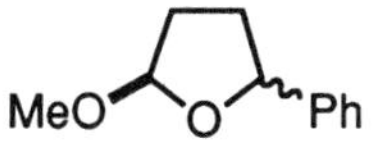 MeO-THF-Ph (cis:trans 50:50)	91
THP-SO₂Ph	THP-thienyl	95
Ph-THP-SO₂Ph	Ph-THP-CH=CH₂	82
AcO, AcO-dihydropyran-SO₂Ph	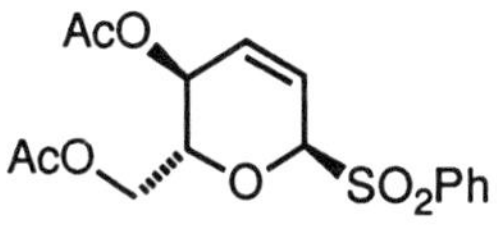 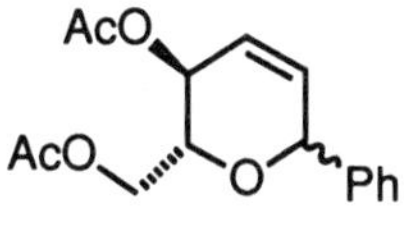 AcO, AcO-dihydropyran-Ph (6R:6S 17:83)	54

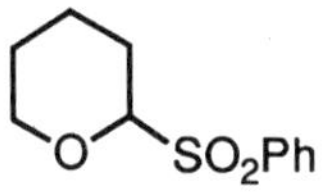

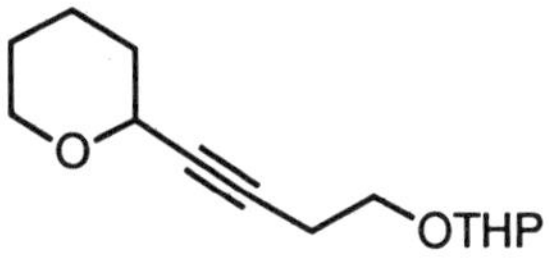

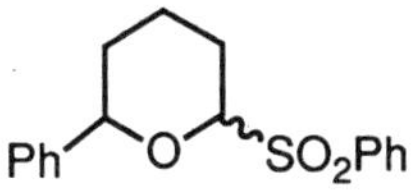

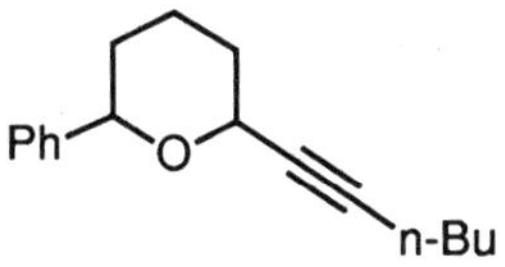

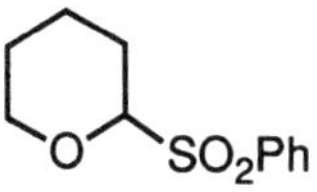

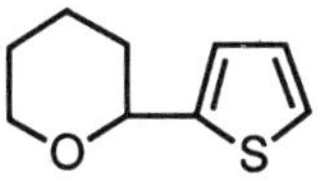

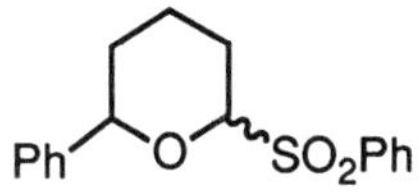

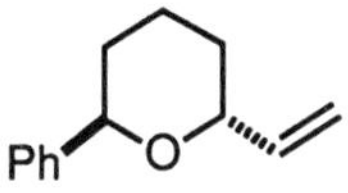

161

The benzenesulfones also undergo nucleophilic displacement with silyl enol ethers and ketene silyl acetals in the presence of a Lewis acid such as aluminum trichloride (Scheme 2).[8]

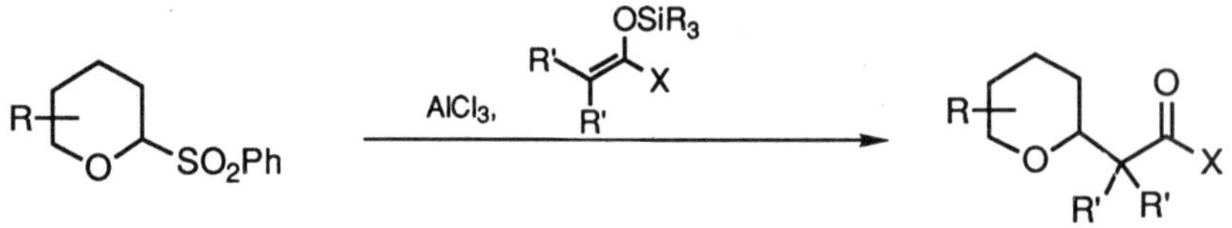

Scheme 2

1. Department of Chemistry, Imperial College of Science, Technology and Medicine, South Kensington, London, SW7 2AY, UK
2. Zelinski, R.; Louvar, J. *J. Org. Chem.* **1958**, *23*, 807.
3. Brown, D. S.; Bruno, M.; Davenport, R. J.; Ley, S. V. *Tetrahedron* **1989**, *45*, 4293.
4. Ley, S. V.; Lygo, B.; Wonnacott, A. *Tetrahedron Lett.* **1985**, *26*, 535; Ley, S. V.; Lygo, B.; Sternfeld, F.; Wonnacott, A. *Tetrahedron* **1986**, *42*, 4333.
5. Greck, C.; Grice, P.; Ley, S. V.; Wonnacott, A. *Tetrahedron Lett.* **1986**, *27*, 5277.
6. Brown, D. S.; Ley, S. V.; Vile, S. *Tetrahedron Lett.* **1988**, *29*, 4873.
7. Brown, D. S.; Ley, S. V. *Tetrahedron Lett.* **1988**, *29*, 4869.
8. Brown, D. S.; Ley, S. V.; Bruno, M. *Heterocycles* **1989**, *28*, 773.

Appendix
Chemical Abstracts Nomenclature (Collective Index Number);
(Registry Number)

Tetrahydro-2-(phenylethynyl)-2H-pyran: 2H-Pyran, tetrahydro-2-(phenylethynyl)- (10); (70141-82-1)

Phenylacetylene: Benzene, ethynyl- (8,9); (536-74-3)

Isopropylmagnesium chloride: Magnesium, chloroisopropyl- (8); Magnesium, chloro(1-methylethyl)- (9); (1068-55-9)

Zinc bromide (8,9); (7699-45-8)

2-(Phenylsulfonyl)tetrahydro-2H-pyran: 2H-Pyran, tetrahydro-2-(phenylsulfonyl)- (11); (96754-03-9)

TRIS(TRIMETHYLSILYL)SILANE

(Trisilane, 1,1,1,3,3,3-hexamethyl-2-(trimethylsilyl)-)

$$Me_3SiCl \;+\; SiCl_4 \;\xrightarrow[THF]{Li}\; (Me_3Si)_4Si$$

$$(Me_3Si)_4Si \;+\; MeLi \;\xrightarrow{\begin{array}{c}1.\;THF\\2.\;H^+/H_2O\end{array}}\; (Me_3Si)_3SiH$$

Submitted by Joachim Dickhaut and Bernd Giese.[1]
Checked by George A. O'Doherty and Leo A. Paquette.

1. Procedure

Lithium powder (7.55 g, 1.07 mol) is placed in a 500-mL, four-necked flask equipped with a condenser, mechanical stirrer, dropping funnel, and low-temperature thermometer (Notes 1 and 2). The apparatus is carefully flushed several times with nitrogen followed by the addition of 50 mL of anhydrous tetrahydrofuran (THF). The reaction flask is cooled to approximately -60°C in a dry ice-acetone bath, and a mixture of freshly distilled (from CaH_2) chlorotrimethylsilane (54.8 mL, 47.1 g, 0.43 mol) and tetrachlorosilane (Note 3) (10.1 mL, 15.0 g, 0.09 mol) in 140 mL of anhydrous THF is added over 1 hr by dropping funnel so that the temperature of the reaction mixture never exceeds -30°C. After addition is complete, stirring is continued for 0.5 hr with cooling (Note 4). The gold-brown suspension is allowed to warm to room temperature and stirred for 12 hr, during which time the color becomes more intense (Note 5). The thermometer is removed and the mixture is heated to reflux for 2 hr to destroy the remaining chlorotrimethylsilane. After the condenser is cooled to room temperature, it is replaced with a nitrogen bubbler and gas inlet. Methyllithium-lithium bromide complex (66 mL, 99 mmol, 1.5 M in ether) is added over 3 hr to the grey-brown mixture with vigorous stirring (Note 6). During the addition a continuous stream of nitrogen is bubbled through the reaction mixture. After the reaction mixture is stirred for an additional 16 hr at room temperature, it acquires a greenish tint. Hydrolysis is carried out by the careful addition of the reaction mixture to 400 mL of ice-cold 2 N hydrochloric acid. [*Caution: The solid residue may be highly pyrophoric. The checkers blanketed the flask with argon prior to the introduction of ether (100 mL) and poured the vigorously stirred slurry into the cold hydrochloric acid. This rinse procedure was repeated twice more.*] The aqueous phase is extracted four times with 200-mL portions of pentane, the combined organic phases are dried over magnesium sulfate and the solvents removed under reduced pressure. Distillation under reduced pressure (1 mm, 38°C) affords 13.4-17.2 g of the product as a clear oil (60-77% yield).

2. Notes

1. The checkers used a three-necked flask having one arm equipped with a Claisen head.

2. All reagents were purchased from Fluka Chemical Corporation, except the methyllithium-lithium bromide complex, which was purchased from Aldrich Chemical Company, Inc., and were used without further purification. The lithium powder can be weighed in air; however, the use of a dust mask is recommended.

3. As tetrachlorosilane smokes strongly when exposed to air, introduction to the addition funnel is best carried out using a syringe.

4. If the temperature falls below -60°C, the mixture may solidify, but returns to a liquid upon warming.

5. The synthesis should be carried out over three days, stirring the mixture overnight. The stirring times reported should be considered a minimum, and need not be followed exactly.

6. A clean dropping funnel should be used for the addition of the methyllithium solution, and should be filled with rigorous exclusion of air. It is simpler to employ a syringe pump, and replace the dropping funnel with a septum. In this case the stream of nitrogen can be introduced by a needle through the septum.

Waste Disposal Information

All toxic materials were disposed of in accordance with "Prudent Practices for Disposal of Chemicals from Laboratories"; National Academy Press; Washington, DC, 1983.

3. Discussion

Tris(trimethylsilyl)silane **1** can be substituted for toxic stannanes like tributylstannane **2** in organic syntheses which involve radicals,[2] because: a) silyl radicals are as efficient as stannyl radicals in the radical-forming step,[3] and b) the Si-H bond strength in tris(trimethylsilyl)silane **1** is only slightly higher than the Sn-H bond strength in tributylstannane **2**.[4]

Bond energy (kcal/mol):

$$\underset{\mathbf{1}}{(Me_3Si)_3Si \overset{79}{\downarrow} H} \qquad \underset{\mathbf{2}}{Bu_3Sn \overset{74}{\downarrow} H}$$

Thus, heating a mixture of an organic bromide or iodide with equimolar amounts of silane **1** and catalytic amounts of a radical initiator like azobisisobutyronitrile gives organic radicals **3** that can undergo addition, cyclization or rearrangement reactions[2] (**3** → **4**) before hydrogen abstraction[5] yields the product.

$$\text{R-Hal} \xrightarrow{(Me_3Si)_3Si \cdot} \underset{\mathbf{3}}{R \cdot} \longrightarrow \underset{\mathbf{4}}{R' \cdot} \xrightarrow[-(Me_3Si)_3Si \cdot]{(Me_3Si)_3SiH} \text{R'-H}$$

Tris(trimethylsilyl)silane **1** is a mediator in this reaction. In contrast to the reported method,[6] the synthesis described in this procedure gives silane **1** in high yields in a one-pot reaction.

1. Institute of Organic Chemistry, University of Basel, St. Johanns-Ring 19, CH-4056 Basel, Switzerland.
2. Giese, B.; Kopping, B.; Chatgilialoglu, C. *Tetrahedron Lett.* **1989**, *30*, 681; Kulicke, K. J.; Giese, B. *Synlett.* **1990**, 91.
3. Ballestri, M.; Chatgilialoglu, C.; Clark, K. B.; Griller, D.; Giese, B.; Kopping, B. *J. Org. Chem.* **1991**, *56,* 678.
4. Kanabus-Kaminska, J. M.; Hawari, J. A.; Griller, D.; Chatgilialoglu, C. *J. Am. Chem. Soc.* **1987**, *109*, 5267.
5. Chatgilialoglu, C.; Griller, D.; Lesage, M. *J. Org. Chem.* **1988**, *53*, 3641.
6. Gilman, H.; Smith, C. L. *J. Organometal. Chem.* **1968**, *14,* 91; Gutekunst, G.; Brook, A. G. *J. Organometal. Chem.* **1982**, *225*, 1; Bürger, H.; Kilian, W. *J. Organomet. Chem.* **1969**, *18*, 299.

Appendix

Chemical Abstracts Nomenclature (Collective Index Number); (Registry Number)

Tris(trimethylsilyl)silane: Trisilane, 1,1,1,3,3,3-hexamethyl-2-(trimethylsilyl)- (8,9); (1873-77-4)

Chlorotrimethylsilane: Silane, chlorotrimethyl- (8,9); (75-77-4)

Tetrachlorosilane: Silicon chloride (8); Silane, tetrachloro- (9); (10026-04-7)

Methyllithium-lithium bromide complex: Lithium, methyl- (8,9); (917-54-4)

9-BORABICYCLO[3.3.1]NONANE DIMER

(9-Borabicyclo[3.3.1]nonane, dimer)

$$\text{cyclooctadiene} \xrightarrow[\text{(CH}_2\text{OMe)}_2]{\text{BH}_3 \cdot \text{SMe}_2} \text{9-BBN dimer}$$

Submitted by John A. Soderquist[1] and Alvin Negron.
Checked by Daniel M. Berger and Larry E. Overman.

1. Procedure

Caution! The manipulation and handling of air-sensitive compounds requires the use of special techniques. While no difficulties have been encountered with the present procedures, the preparer should consult References 2 and 3 prior to carrying out these syntheses.

A 2-L, three-necked, round-bottomed flask containing a magnetic stirring bar is fitted with a 250-mL addition funnel and a distillation assembly set for downward distillation to a 500-mL receiver flask. Rubber septa are used to isolate the system from atmospheric contact. Under a nitrogen purge, vented to an exhaust hood through a mercury bubbler, the entire system is thoroughly flame-dried (Note 1). After the 2-L flask is cooled to room temperature, it is charged with 500 mL of pure, dry 1,2-dimethoxyethane (Note 2) and 153 mL (1.53 mol) of borane-methyl sulfide complex (Note 3) employing a double-ended needle to effect the transfer. With a similar technique, 164 g (1.52 mol) of 1,5-cyclooctadiene (Note 4) is transferred to the addition funnel. To the stirred borane solution, 1,5-cyclooctadiene is added dropwise over ca. 1 hr to maintain a reaction temperature of 50-60°C during which time a small amount of dimethyl sulfide (bp 38°C) distills slowly from the reaction mixture. After the addition is completed, the addition funnel is replaced with a glass stopper and approximately 300 mL of the solution is distilled to reach a final distillation temperature of 85°C, indicating the complete removal of dimethyl sulfide from the reaction mixture (Note 5). If the distillate temperature does not reach 85°C, 150 mL of additional 1,2-dimethoxyethane is added and the distillation is continued until the distillate temperature reaches 85°C. The distillation assembly is replaced with a rubber septum and 1,2-dimethoxyethane is added to the reaction flask to bring the total liquid volume to 1 L. The mixture is warmed to effect the dissolution of the solid and allowed to cool very slowly to 0°C, which results in the formation of crystalline 9-borabicyclo[3.3.1]nonane (9-BBN) dimer. The supernatant liquid is decanted from the product using a double-ended needle and the 9-BBN dimer is dissolved in 1 L of fresh 1,2-dimethoxyethane. After the flask is cooled to 0°C, the supernatant liquid is removed as above and the large needles are dried under reduced pressure for 12 hr at 0.1 mm to give 158-165 g (85-89%) of product (mp 152-154°C, sealed capillary) (Notes 6-8).

2. Notes

1. Alternatively, the apparatus can be dried for 4 hr at 150°C, assembled hot and purged with dry nitrogen.

2. 1,2-Dimethoxyethane, available from the Aldrich Chemical Company, Inc., was predried over calcium hydride and distilled from sodium/benzophenone prior to use. The solvent was used directly after purification or stored in an ampule bottle, available from the Aldrich Chemical Company, Inc., under a nitrogen atmosphere.

3. Borane-methyl sulfide complex, obtained from the Aldrich Chemical Company, Inc., was used directly without additional purification. However, titration of

the reagent was carried out with glycerol as described[2] to determine its actual molarity. Older samples of this reagent can be distilled under aspirator vacuum to obtain pure reagent.

4. 1,5-Cyclooctadiene, obtained from the Aldrich Chemical Company, Inc., was distilled under aspirator pressure from lithium aluminum hydride prior to use.

5. Failure to remove the dimethyl sulfide from the reaction mixture increases the solubility of the 9-BBN dimer and lowers the overall yield to ca. 65%.

6. The spectra of the product are as follows: ^{1}H NMR (300 MHz, C_6D_6) δ: 1.44-1.57 (m, 4 H), 1.58-1.74 (m, 12 H), 1.83-2.07 (m, 12 H). A standard HETCOR experiment revealed that protons on each of the methylene carbons were superimposed upon one another to give rise to these downfield multiplets; ^{13}C NMR (75 MHz, C_6D_6) δ: 20.2 (br, C-1,5), 24.3 (C-3,7), 33.6 (C-2,4,6,8); ^{11}B NMR (96 MHz, C_6D_6) δ: 28.

7. The 9-BBN dimer so prepared is reasonably air-stable so that exposure to the atmosphere for 1 month lowered the mp to ca. 146-151°C.

8. Purification of commercial 9-BBN and other samples can be effected by recrystallization from 1,2-dimethoxyethane. Insoluble impurities can be removed from hot 1,2-dimethoxyethane solutions of 9-BBN by decantation of the solution to a second dry flask. To prevent clogging of the double-ended needle during the transfer process it is important to keep the ends of needle below the liquid surfaces. We have found that the receiver vessel should be charged with a small quantity of fresh, hot 1,2-dimethoxyethane prior to decantation and that a portion of this material should be transferred under a positive pressure of nitrogen to the 9-BBN solution to warm initially the transfer needle. Subsequently, the hot 9-BBN solution can be transferred without difficulty.

Waste Disposal Information

All toxic materials were disposed of in accordance with "Prudent Practices for Disposal of Chemicals from Laboratories"; National Academy Press; Washington, DC, 1983.

3. Discussion

9-Borabicyclo[3.3.1]nonane (9-BBN) has been prepared by the thermal redistribution of 9-n-propyl-9-BBN,[4] and the hydroboration of 1,5-cyclooctadiene with borane-tetrahydrofuran complex followed by thermal isomerization of the mixture of dialkylboranes at 65°C.[5] Solutions of 9-BBN have been prepared from the hydroboration of 1,5-cyclooctadiene with borane-methyl sulfide in solvents other than THF.[6] The present procedure involves the cyclic hydroboration of 1,5-cyclooctadiene with borane-methyl sulfide in 1,2-dimethoxyethane.[7] Distillative removal of the dimethyl sulfide in this special solvent system provides a medium that gives high purity, large needles of crystalline 9-BBN dimer in excellent yield. The material can be handled in air for brief periods without measurable decomposition.

As a dialkylborane, 9-borabicyclo[3.3.1]nonane (9-BBN) is unrivaled in both stability and selectivity.[8] It has been distilled (bp 195°C, 12 mm) and exhibits a strong characteristic IR absorption band at 1560 cm^{-1} (B-H-B) for the bridged dimeric structure.[5] The crystal structure of 9-BBN dimer has been determined[9] and the drawing above approximates the conformational features of this compound. The ^{13}C NMR properties of 9-BBN adducts have been studied extensively.[10]

Since the 9-methoxy derivative of 9-BBN is a common by-product of several reactions of 9-BBN,[11] its efficient conversion back to 9-BBN has been described.[12] Such a process enables one to recycle 9-BBN in reactions which require its high regioselectivity in hydroboration reactions and the related organoborane conversions.

The selective transformations of 9-BBN are numerous and varied, with derivatives being readily prepared through both hydroboration and organometallic methodology.[8] It has been used for the preparation of isomerically-pure boracycles,[11,13] the highly enantioselective reduction of aldehydes and ketones,[14] the preparation of new selective borohydride reducing agents,[15] C-C bond-forming transformations,[16] and radiopharmaceutical labeling.[17] Its reactivity has made it the reagent of choice for many organoborane conversions.[18] The stability and distinctive spectral properties of 9-BBN have provided the initial key information to unravel the details of hydroboration reactions.[8,19]

1. Department of Chemistry, University of Puerto Rico, Rio Piedras, PR 00931.
2. Brown, H. C.; Kramer, G. W; Levy, A. B.; Midland, M. M. "Organic Syntheses Via Boranes"; Wiley-Interscience: New York, 1975.
3. "Handling of Air-Sensitive Reagents"; Aldrich Technical Product Bulletin No. AL-134, 1983.
4. Köster, R. *Angew. Chem.* **1960**, *72*, 626.
5. Knights, E. F.; Brown, H. C. *J. Am. Chem. Soc.* **1968**, *90*, 5280; Brown, H. C.; Knights, E. F.; Scouten, C. G. *J. Am. Chem. Soc.* **1974**, *96*, 7765.
6. Brown, H. C.; Mandal, A. K.; Kulkarni, S. U. *J. Org. Chem.* **1977**, *42*, 1392.
7. Soderquist, J. A.; Brown, H. C. *J. Org. Chem.* **1981**, *46*, 4599.
8. Brown, H. C.; Lane, C. F. *Heterocycles* **1977**, *7*, 453; Rao, V. V. R.; Mehrotra, I.; Devaprabhakara, D. *J. Sci. Ind. Res.* **1979**, *38*, 368; Zaidlewicz, M. In "Comprehensive Organometallic Chemistry"; Wilkinson, G.; Stone, F. G. A.;

Abel, E. W., Eds.; Pergamon Press: Oxford, 1982; Vol. 7, pp 161 and 199; Pelter, A.; Smith, K.; Brown, H. C. "Borane Reagents"; Academic Press: London, 1988.

9. Brauer, D. J.; Krueger, C. *Acta Crystallogr., Sect. B* **1973**, *29*, 1684.

10. Brown, H. C.; Soderquist, J. A. *J. Org. Chem.* **1980**, *45*, 846; Blue, C. D.; Nelson, D. J. *J. Org. Chem.* **1983**, *48*, 4538; Soderquist, J. A.; Colberg, J. C.; Del Valle, L. *J. Am. Chem. Soc.* **1989**, *111*, 4873; Soderquist, J. A.; Rivera, I.; Negron, A. *J. Org. Chem.* **1989**, *54*, 4051.

11. Soderquist, J. A.; Shiau, F.-Y.; Lemesh, R. A. *J. Org. Chem.* **1984**, *49*, 2565; Soderquist, J. A.; Negron, A. *J. Org. Chem.* **1989**, *54*, 2462 and references cited therein.

12. Soderquist, J. A.; Negron, A. *J. Org. Chem.* **1987**, *52*, 3441; Brown, H. C.; Kulkarni, S. U. *J. Organomet. Chem.* **1979**, *168*, 281.

13. Brown, H. C.; Pai, G. G. *J. Organomet. Chem.* **1983**, *250*, 13; Soderquist, J. A.; Najafi, M. R. *J. Org. Chem.* **1986**, *51*, 1330.

14. Midland, M. M.; Graham, R. S. *Org. Synth.* **1985**, *63*, 57; Brown, H. C.; Jadhav, P. K. In "Asymmetric Synthesis"; Morrison, J. D., Ed.; Academic Press: New York, 1983; Vol. 2, Chapter 1; Midland, M. M. In "Asymmetric Synthesis"; Morrison, J. D., Ed.; Academic Press: New York, 1983; Vol. 2, Chapter 2; Midland, M. M.; McLoughlin, J. I.; Gabriel, J. *J. Org. Chem.* **1989**, *54*, 159.

15. Brown, H. C.; Park, W. S.; Cho, B. T. *J. Org. Chem.* **1986**, *51*, 1934; Brown, H. C.; Cho, B. T.; Park, W. S. *J. Org. Chem.* **1988**, *53*, 1231; Narasimhan, S. *Indian J. Chem., Sect. B* **1986**, *25B*, 847; Cha, J. S.; Yoon, M. S.; Kim, Y. S.; Lee, K. W. *Tetrahedron Lett.* **1988**, *29*, 1069; Soderquist, J. A.; Rivera, I. *Tetrahedron Lett.* **1988**, *29*, 3195; Cha, J. S.; Lee, K. W.; Yoon, M. S.; Lee, J. C.; Yoon, N. M. *Heterocycles* **1988**, *27*, 1713.

16. Miyaura, N.; Ishiyama, T.; Sasaki, H.; Ishikawa, M.; Satoh, M.; Suzuki, A. *J. Am. Chem. Soc.* **1989**, *111*, 314; Hooz, J.; Oudenes, J.; Roberts, J. L.; Benderly, A. *J. Org. Chem.* **1987**, *52*, 1347; Soderquist, J. A.; Santiago, B.; Rivera, I. *Tetrahedron Lett.* **1990**, *31*, 4981; Soderquist, J. A.; Santiago, B. *Tetrahedron Lett.* **1990**, *31*, 5541.

17. Kothari, P. J.; Finn, R. D.; Vora, M. M.; Boothe, T. E.; Emran, A. M.; Kabalka, G. W. *Int. J. Appl. Radiat. Isot.* **1985**, *36*, 412; Kabalka, G. W.; Delgado, M. C.; Kunda, U. S.; Kunda, S. A. *J. Org. Chem.* **1984**, *49*, 174.

18. For example, see: Soderquist, J. A.; Hassner, A. *J. Organomet. Chem.* **1978**, *156*, C12; Soderquist, J. A.; Brown, H. C. *J. Org. Chem.* **1980**, *45*, 3571; Brown, H. C.; Molander, G. A.; Singh,. S. M.; Racherla, U. S. *J. Org. Chem.* **1985**, *50*, 1577; Brown, H. C.; Vara, Prasad, J. V. N.; Zee, S.-H. *J. Org. Chem.* **1985**, *50*, 1582; Brown, H. C.; Cha, J. S.; Nazer, B.; Brown, C. A. *J. Org. Chem.* **1985**, *50*, 549; Molander, G. A.; Singaram, B.; Brown, H. C. *J. Org. Chem.* **1984**, *49*, 5024; Brown, H. C.; Narasimhan, S. *J. Org. Chem.* **1984**, *49*, 3891; Brown, H. C.; Mathew, C. P.; Pyun, C.; Son, J. C.; Yoon, N. M. *J. Org. Chem.* **1984**, *49*, 3091; Yamataka, H.; Hanafusa, T. *J. Org. Chem.* **1988**, *53*, 772; Bubnov, Yu. N.; Zheludeva, V. I. *Izv. Akad. Nauk SSSR., Ser. Khim.* **1987**, 235; *Chem. Abst.* **1987**, *107*, 197578b; Brown, H. C.; Midland, M. M.; Kabalka, G. W. *Tetrahedron* **1986**, *42*, 5523; Liu, C.; Wang, K. K. *J. Org. Chem.* **1986**, *51*, 4733; Fleming, I.; Lawrence, N. J. *Tetrahedron Lett.* **1988**, *29*, 2073, 2077; Köster, R.; Schüssler, W.; Yalpani, M. *Chem. Ber.* **1989**, *122*, 677.

19. Brown, H, C.; Wang, K. K.; Scouten, C. G. *Proc. Nat. Acad. Sci. USA* **1980**, *77*, 698; Nelson, D. J.; Cooper, P. J.; Coerver, J. M. *Tetrahedron Lett.* **1987**, *28*, 943; Nelson, D. J.; Cooper, P. J. *Tetrahedron Lett.* **1986**, *27*, 4693.

Appendix

Chemical Abstracts Nomenclature (Collective Index Number); (Registry Number)

9-Borabicyclo[3.3.1]nonane dimer: 9-Borabicyclo[3.3.1]nonane, dimer (10); (70658-61-6); Diborane (6), 1,1:2,2-di-1,5-cyclooctylene- [Available from Aldrich Chemical Company, Inc.] (8,9); (21205-91-4)

9-Borabicyclo[3.3.1]nonane (8,9); (280-64-8)

Dimethoxyethane: Ethane, 1,2-dimethoxy- (8,9); (110-71-4)

Borane-methyl sulfide complex: Methyl sulfide, compd. with borane (1:1) (8); Borane, compd. with thiobis[methane] (1:1) (9); (13292-87-0)

1,5-Cyclooctadiene (8,9); (111-78-4)

9-Methoxy-9-borabicyclo[3.3.1]nonane: 9-Borabicyclo[3.3.1]nonane, 9-methoxy- (9); (38050-71-4)

CYCLOPROPANATION USING AN IRON-CONTAINING METHYLENE TRANSFER REAGENT: 1,1-DIPHENYLCYCLOPROPANE

(Iron (1+), dicarbonyl(η^5-2,4-cyclopentadien-1-yl)(dimethylsulfonium η-methylide)-, tetrafluoroborate (1-))

$$[\text{Cp-Fe(CO)}_2]_2 \xrightarrow[\substack{(3)\ CH_3I \\ (4)\ NaBF_4/H_2O}]{\substack{(1)\ \text{Na dispersion/THF} \\ (2)\ ClCH_2SCH_3}} \text{Cp-}\underset{CO}{\overset{OC}{\text{Fe}}}\text{-CH}_2\text{S}^+(CH_3)_2\ BF_4^-$$

$$\text{Cp-}\underset{CO}{\overset{OC}{\text{Fe}}}\text{-CH}_2\text{S}^+(CH_3)_2\ BF_4^- \xrightarrow[\text{dioxane}]{\underset{Ph}{\overset{Ph}{>}}C=CH_2} \underset{\underset{Ph}{Ph}}{\overset{H\ \ H}{\underset{C}{\triangle}}}\underset{H}{\overset{H}{C}}$$

Submitted by Matthew N. Mattson,[1a] Edward J. O'Connor,[1b] and Paul Helquist.[1a]
Checked by Jörn-Bernd Pannek and Ekkehard Winterfeldt.

1. Procedure

CAUTION! This experiment should be performed in an efficient fume hood because of the unpleasant odors of sulfide-containing materials. In addition, the first part of this procedure should be conducted behind a safety shield because of the use of highly reactive sodium metal.

Into a dry, one-necked, 2000-mL, round-bottomed flask is placed a medium-sized magnetic stirring bar (Note 1) and cyclopentadienyliron dicarbonyl dimer [$C_5H_5(CO)_2Fe$]$_2$, (0.50 mol equiv, 0.21 mol, 74.4 g; Notes 2 and 3). Sodium dispersion (40% by weight) in light mineral oil (1.25 mol equiv, 0.52 mol, 30.1 g; Notes

4 and 5) is weighed into the flask (Note 6). The flask is then equipped with a reflux condenser topped with a three-way stopcock (Note 7) having a vertical tubulation capped with a septum through which solvents and reagents can be introduced with long needles or cannulas. By evacuation through the other tubulation of the stopcock, the apparatus is evacuated and filled with nitrogen twice, then placed under vacuum ($\leq$ 0.1 mm) for 1 to 2 hr to remove the bulk of the mineral oil. The flask is filled with nitrogen, and tetrahydrofuran (THF; 850 mL; Note 8) is transferred into the flask. Rapid stirring is begun and maintained while an oil bath or a heating mantle is employed to heat the mixture at reflux for $\geq$ 18 hr.

The flask is cooled to 0°C in an ice bath, and chloromethyl methyl sulfide (1.00 mol equiv, 0.42 mol, 35.2 ml) is added dropwise with a syringe over 25 min (Notes 9 and 10). After residues of the sulfide are rinsed into the flask with additional THF (ca. 5-10 mL), the mixture is stirred at 0°C for 1 hr and then at 25°C for 1 hr (Note 11). Iodomethane (1.30 mol equiv, 0.55 mol, 34.0 mL; Note 12) is added over 5 min using a syringe. After residues of iodomethane are rinsed into the flask with THF (5-10 mL), the mixture is stirred at 25°C for $\geq$ 15 hr. Stirring is stopped (Note 13), and the volatile materials are removed under vacuum ($\leq$ 0.1 mm) using a large, liquid nitrogen-cooled trap (Note 14). The vacuum in the apparatus is relieved with nitrogen, and the three-way stopcock is removed from the top of the condenser, exposing the reaction mixture to air.

In a 2000-mL Erlenmeyer flask containing a magnetic stirring bar, a solution of sodium tetrafluoroborate (6.00 mol equiv, 2.52 mol, 277 g) in water (1200 mL total volume of solution) is prepared and heated to 95°C while being stirred. A 1000-mL portion of the hot sodium tetrafluoroborate solution solution is slowly poured down the condenser into the reaction mixture which is kept at ca. 95°C while being stirred. At the same time, a 350-mL, medium-frit, sintered-glass Büchner funnel is prepared with a 2.5-cm layer of diatomaceous earth and a 1-cm layer of sand covered with a piece of filter paper with holes punched in it, and the funnel is preheated by passage, with suction, of 700-1000 mL of hot, distilled water which is then discarded. The condenser is removed from the reaction flask, and the contents are suction-filtered through the hot funnel into a heated, 2000-mL filter flask (Note 15). The remaining hot sodium tetrafluoroborate solution is used to rinse the reaction flask and the hot funnel. The combined filtrates are swirled while being cooled. If necessary, a seed crystal can be added. The filtration flask is placed in an ice bath while swirling is continued. After the temperature reaches 0°C, the flask is placed in a freezer at ca. -10°C for 1-3 hr. The product is collected by suction filtration using a large, chilled Büchner funnel (Whatman no. 1 filter paper) and is rinsed with ice-cold distilled water (150 mL) and cold diethyl ether (1500 mL). The filter cake is broken up, and the crystals are dried in a stream of air overnight. There is obtained 100.6 g (70.4%) of (η^5-C_5H_5)(CO)$_2$FeCH$_2$S$^+$(CH$_3$)$_2$ BF$_4^-$ as free-flowing, flake-like, amber crystals (Notes 16-18). The yields were found to be considerably lower on runs of smaller scale (Note 19).

Into a 200-mL, one-necked, round-bottomed flask equipped with a magnetic stirring bar are placed the crystalline reagent (35 g, 0.10 mol; Note 20), 1,1-diphenylethene (9.1 mL, 9.3 g, 0.05 mol; Note 21), and dioxane (25 mL; Notes 22 and 23). The flask is equipped with a reflux condenser topped with a stopcock, and a nitrogen atmosphere (Note 24) is established within the apparatus. While being stirred vigorously, the heterogeneous mixture is heated to reflux in an oil bath (120°C) for 14 hr (Note 25). The brown mixture is removed from the oil bath and allowed to cool sufficiently to permit the addition of hexane (75 mL, Note 26) to the flask. The mixture is stirred in the air until the flask reaches 25°C. The supernatant liquid containing the product is poured from the flask and filtered through Whatman no. 1 filter paper. The remaining solid is repeatedly suspended and washed with several portions of hexane (ca. 1000 mL total; Note 27). The combined filtrates are filtered through a pad of silica gel in a sintered glass Büchner funnel and are then concentrated by rotary evaporation. The residual dark brown oil is dissolved in methanol (200 mL) to give an orange-brown solution which immediately becomes dark green when solid ferric chloride (7 g; Note 28) is added at 25°C. The mixture is stirred for 15 min and then concentrated by rotary evaporation. The residual dark green oil is extracted with two 200-mL portions of hexane, and the combined extracts are filtered through a pad of silica gel and concentrated by rotary evaporation. The colorless oil that remains is distilled through a short-path apparatus to give 8.76 g (88%) of 1,1-diphenylcyclopropane as a clear, colorless liquid, bp 89°C (0.8 mm; lit[8] 110-111°C, 1.3 mm; Note 29). The checkers obtained 65-77% yield of product on roughly half the scale.

2. Notes

1. The stirring bar must be able to stir the heterogeneous reaction mixture rapidly. Very good stirring is required for the metallic sodium dispersion to react efficiently. A medium-sized, egg-shaped stirring bar (32 x 16 mm, available from Fisher Scientific Company) was found to be particularly effective.

2. Cyclopentadienyliron dicarbonyl dimer $[C_5H_5(CO)_2Fe]_2$ can be purchased from Alfa Products, Morton/Thiokol Inc. or Aldrich Chemical Company, Inc. Alternatively, it is easily and inexpensively prepared by heating dicyclopentadiene with iron pentacarbonyl. Our yield (80-90%) of this reagent is considerably higher than that reported in the literature procedure.[2]

3. In order to allow for proper placement of the sodium dispersion in the flask later (Note 5), the $[C_5H_5(CO)_2Fe]_2$ was neatly piled in a mound on top of the stirring bar in the middle of the bottom of the flask.

4. The 40% (by weight) sodium dispersion in light mineral oil was used as obtained from Aldrich Chemical Company, Inc., except for thorough shaking immediately prior to transfer of the dispersion.

5. A 1-cm diameter glass tube narrowed to a tip at one end and equipped with a pipet bulb was used to transfer the dispersion which was carefully placed around the perimeter of the mound of $[C_5H_5(CO)_2Fe]_2$. After evaporation of the oil, the stirring bar should rest in the center of the ring of sodium without contacting it. In this way, the reaction mixture can subsequently be stirred more efficiently. Also, all of the sodium should lie below the surface of the tetrahydrofuran solution formed, so that the mixture reacts efficiently upon being heated at reflux.

6. The procedure described here for reductive cleavage of this compound with sodium dispersion[3] to give sodium cyclopentadienyldicarbonylferrate is considerably more convenient and less hazardous than the more traditional use of sodium amalgam that was reported previously.[4]

7. The stopcock used in this procedure is of the design shown in Figure 1. A source of inert gas and vacuum can be attached to the horizontal tubulation. The vertical tubulation is capped with a septum to allow introduction of liquid reagents and solvents through use of a long syringe needle or cannula inserted through the septum and down through the body of the stopcock. In order to avoid air leaks through the septum into the reaction apparatus when reagents are not being added, the stopcock is normally turned to close off the vertical tubulation, but to leave the flask open to the nitrogen/vacuum source.

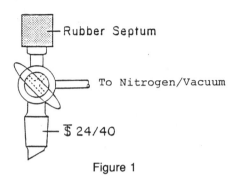

Figure 1

8. Commercial, anhydrous-grade tetrahydrofuran (THF) is further purified by distillation from a dark blue or purple solution of sodium benzophenone ketyl or dianion under nitrogen. One method for transferring the THF into the reaction flask is through the use of cannulas. The cannulas (available from Aldrich Chemical Company, Inc.) are constructed from 60-cm sections of 18-gauge stainless steel tubing with a needle tip at each end. It is perhaps more convenient to use two short sections of needle tubing (each having a needle point at only one end) joined with 25-50 cm of small-diameter Teflon tubing. Transfer through the cannula is facilitated by applying a slight vacuum to the reaction apparatus while maintaining a positive pressure of nitrogen in the flask originally containing the distilled THF. Alternatively, the THF can be distilled directly into the reaction flask.

9. Chloromethyl methyl sulfide was obtained from Aldrich Chemical Company, Inc. and distilled under nitrogen prior to use, although direct use of the commercial material without distillation had little effect on the overall efficiency of this procedure.

10. *WARNING:* Chloromethyl methyl sulfide has a very unpleasant, penetrating odor and should be handled in a properly ventilated fume hood. Also, because of its structural similarity to chloromethyl methyl ether which is highly toxic and an OSHA-regulated carcinogen, this sulfide should be handled as a substance having potentially similar toxic properties.

11. The product of this alkylation step is $(\eta^5\text{-}C_5H_5)(CO)_2FeCH_2SCH_3$ which is used directly in the next step but which, if desired, can be isolated as a dark yellow-brown, somewhat air-sensitive oil in greater than 90% yield.[5,6]

12. Iodomethane (99%) was used as obtained from Aldrich Chemical Company, Inc. Excess iodomethane is used to quench any unreacted sodium metal.

13. At this point, the flask can be swirled so that any small amounts of sodium adhering to the wall of the flask above the solution level can be coated with the reaction mixture.

14. Trapping of the unreacted chloromethyl methyl sulfide in the cold trap is recommended because of the problems summarized in Note 10.

15. In order for this filtration to proceed smoothly, the funnel and its contents must remain hot to avoid premature crystallization of the product and clogging of the funnel. Minor clogging can be remedied by addition of a 100-mL portion of boiling distilled water to the funnel. Major clogging may require addition of boiling water and agitation of the filtration media with a spatula. This addition, however, may reduce the yield of the crystallized product.

16. Physical data for this compound are the following: mp 129-130°C (corrected); IR (KBr pellet) cm^{-1}: 3120, 3035, 2040, 1955, 1417, 1328, 1280, 1055, 852; ^{1}H NMR (80 MHz, CD$_3$NO$_2$) δ: 2.72 (s, 2 H, C$\underline{H}_2$), 3.00 (s, 6 H, 2 C$\underline{H}_3$), 5.34 (s, 5 H, C$_5\underline{H}_5$); ^{13}C NMR (20 MHz, CD$_3$NO$_2$) 13.35 ($\underline{C}H_2$), 31.20 ($\underline{C}H_3$), 87.88 ($\underline{C}_5H_5$), 215.56 ($\underline{C}O$).

17. This material is satisfactory for alkene cyclopropanation reactions, although recrystallization can be effected very easily by dissolving the crude product in nitromethane at 25°C in the air and by slowly cooling the filtered solution to -70°C. The recrystallization recovery is greater than 80% and provides large, "gem-like," amber-colored crystals. Acetone can also be used as the recrystallization solvent.

18. This reagent can be stored in ordinary flasks or bottles in the air, but it should be protected from bright light, which leads to slow decomposition. Storage in a dark brown bottle is recommended.

19. The checkers' yields ranged from 25-46% in preparations that were run on one-fourth to one-half of the scale used by the submitters.

20. A two-fold excess of the iron reagent is employed to assure high conversion of the alkene to the cyclopropane. Equimolar amounts of the starting materials can be used, but the cyclopropane yield is ca. 20% lower.

21. 1,1-Diphenylethene is obtained from Aldrich Chemical Company, Inc. and is used without further purification.

22. 1,4-Dioxane is distilled from sodium benzophenone ketyl under nitrogen.

23. Nitromethane is also a good solvent for this reaction, and in some cases gives somewhat higher yields of cyclopropanes. Also, the reaction times are reduced to 2-4 hr when nitromethane is used. Before use, this solvent is purified according to a published procedure.[7] Commercially obtained solvent is first dried over anhydrous magnesium sulfate and then over anhydrous calcium sulfate. The solvent is filtered into a flask containing activated 3 Å molecular sieves and is heated at 60°C for 8 hr while being stirred. Nitromethane is distilled from the powdered molecular sieves under reduced pressure (bp 58°C, 150 mm; lit.[7] 58°C, 160 mm) directly into a flask containing additional 3 Å molecular sieves. The purified solvent is stored in the dark. When "wet" nitromethane from commercial sources is used directly as the reaction solvent, the percent conversions of alkenes to cyclopropanes are reduced substantially. *CAUTION*: Distillations of nitromethane and reactions using this solvent at elevated temperature should be conducted behind a safety shield.

24. When the cyclopropanation reactions are run in the presence of air, the yields are slightly reduced.

25. Vigorous stirring is necessary for a reasonable rate of reaction. The mixture remains heterogeneous both before and after the iron reagent melts. Monitoring of the reaction by GLPC (2-m 5% OV-1 or SE-30) is recommended to assure maximum conversion before the reaction is stopped.

26. The function of the hexane (or pentane) is to promote precipitation of organometallic byproducts.

27. The solid is bright yellow after these washings and consists primarily of $[C_5H_5(CO)_2FeS(CH_3)_2]^+$ BF_4^- and some unreacted cyclopropanation reagent. The latter can be recovered if desired by recrystallization of this mixture from acetone.

28. Ferric chloride destroys ferrocene, a contaminating side product that is difficult to remove by physical means because of its hydrocarbon-like characteristics.

29. The purity of this product is greater than 98% as determined by GLPC (2-m 5% OV-1 or SE-30). The spectral properties are as follows: ^{1}H NMR (300 MHz, CDCl$_3$, cf. lit.[8]) δ: 1.30 (s, 4 H, 2 cyclopropyl CH$_2$), 7.12-7.40 (m, 10 H, ArH); ^{13}C NMR (75 MHz, CDCl$_3$) δ: 16.33 (cyclopropyl CH$_2$), 29.97 (quaternary cyclopropyl C), 125.9 (para C), 128.2, 128.4 (ortho and meta C), 145.8 (ipso C); MS (EI, 70 eV) m/e (rel intensity) 194 (M$^+$, 86), 193 (100), 178 (64), 115 (9).

Waste Disposal Information

All toxic materials were disposed of in accordance with "Prudent Practices for Disposal of Chemicals from Laboratories"; National Academy Press; Washington, DC, 1983.

3. Discussion

Despite their high ring strain, cyclopropanes are commonly encountered among naturally occurring as well as synthetic compounds. Cyclopropanes are most commonly synthesized by addition of alkylidene units to alkenes.[9] These reactions employ various types of carbenes, carbenoids, or diazo compounds. One particularly important method is the Simmons-Smith reaction[9b] which, according to the original procedure, involves the treatment of diiodomethane with zinc-copper couple to generate a reactive intermediate that serves as a cyclopropanation reagent. Several modifications of this procedure have been reported more recently. Another common approach is to employ a dihalocarbene to give a 1,1-dihalocyclopropane which is treated subsequently with a reducing agent to effect replacement of the halide substituents by hydrogen.

In the mid-1960's, transition metal carbene complexes were first reported by E. O. Fischer.[10] Although their structures may be suggestive of classical carbene-like behavior, relatively few of these complexes serve as useful cyclopropanation reagents.[11] Rather, these compounds exhibit their own characteristic types of reactions, many of which are useful in synthetic transformations other than cyclopropanations. Contrary to this more general case, Pettit[12] and then Green[13] reported some early findings that indicated the possible utility of certain iron carbene complexes for three-membered ring construction. Their studies were followed by investigations of iron complexes by many others,[9m,14] among which has been the recent work of Brookhart[15] and Casey.[16]

Helquist has focused efforts on developing synthetically useful cyclopropanation reagents based upon the use of stable organoiron compounds which may be regarded, at least formally, as direct precursors of reactive carbene complexes. Sulfonium derivatives[6b,17] and alkenyl complexes[18] have proven to be useful in this regard. The presently described methylene transfer reagent[6b,17a,b] and a related ethylidene transfer reagent[17c,d] are included among the former sulfonium salt complexes. Among the many compounds reported by Brookhart is a useful silyl ether-based reagent for ethylidene transfer[15a] as well as a reagent for asymmetric cyclopropanations.[15b] Helquist[18] and Casey[16] have also reported complexes for transfer of several more complex types of alkylidene units.

An important advantage of the presently described procedure is that the cyclopropanation reagent is unusually stable for an organometallic compound. Not only is the solid reagent stable to air indefinitely, but its crystallization is accomplished from hot aqueous solutions. Samples of this reagent have been stored in ordinary laboratory reagent bottles for more than five years with no noticeable decomposition. This stability is in contrast to typical Simmons-Smith intermediates and diazoalkanes. Another advantage of the present reagent is that once it has been prepared, its subsequent use in cyclopropanation reactions is trivially straightforward. The reagent can be handled as an ordinary laboratory reagent and combined with an alkene substrate and a suitable solvent in an ordinary flask. Although an inert atmosphere is specified for the present cyclopropanation, these reactions have also been performed routinely in the air with only small reductions in yields.

1,1-Diphenylcyclopropane has been prepared previously by (1) the Simmons-Smith procedure (24% yield)[9b,19] and modified versions of this method (up to 72%),[20] (2) sulfonium ylide addition to 1,1-diphenylethene (61% yield),[21] (3) reduction of 1,1-diphenyl-2,2-dihalocyclopropanes with sodium in ammonia (47% yield),[22] with sodium and tert-butyl alcohol (80%),[8] or with diethyl lithiomethanephosphonate (62%),[23] (4) base-promoted cyclization of trimethyl(3,3-diphenylpropyl)ammonium iodide (78%),[24] (5) boron trifluoride-promoted cyclization of a corresponding 3-hydroxypropylstannane (97%),[25] (6) reaction of 3,3-diphenylpropenoic acid with lithium aluminum hydride (62%),[26] (7) reaction of diphenylmethane with $NaAlH_2(OCH_2CH_2OCH_3)_2$ (20%),[27] (8) decomposition of the pyrazoline (unspecified yield) obtained by addition of diazomethane to 1,1-diphenylethene,[28] (9) thermolysis of a corresponding acyldiazene (43%),[29] (10) photolysis of 2,2-diphenylcyclobutanone (19%),[29] and (11) decarboxylation of 1,1-diphenyl-2-carboxycyclopropane (unspecified yield).[30]

The Table summarizes some of the other examples of cyclopropanations that have been performed by the presently described procedure.[6b,17a,c,31]

1. (a) University of Notre Dame; (b) State University of New York at Stony Brook,
2. King, R. B.; Stone, F. G. A. *Inorg. Synth.* **1963**, *7,* 99.
3. Reger, D. L.; Fauth, D. J.; Dukes, M. D. *Synth. React. Inorg. Metal-Org. Chem.* **1977**, *7,* 151.
4. (a) King, R. B.; Bisnette, M. B. *J. Organomet. Chem.* **1964**, *2,* 15; (b) Ellis, J. E.; Faltynek, R. A.; Hentges, S. G. *J. Organomet. Chem.* **1976**, *120,* 389; (c) Gladysz, J. A.; Williams, G. M.; Tam, W.; Johnson, D. L. *J. Organomet. Chem.* **1977**, *140,* C1.
5. King, R. B.; Bisnette, M. B. *Inorg. Chem.* **1965**, *4,* 486.
6. (a) O'Connor, E. J. Ph.D. Dissertation, State University of New York at Stony Brook, 1984; (b) O'Connor, E. J.; Brandt, S.; Helquist, P. *J. Am. Chem. Soc.* **1987**, *109,* 3739.
7. Smith, S. G.; Fainberg, A. H.; Winstein, S. *J. Am. Chem. Soc.* **1961**, *83,* 618.
8. Adam, W.; Birke, A.; Cadiz, C.; Diaz, S.; Rodriguez, A. *J. Org. Chem.* **1978**, *43,* 1154.
9. For some reviews of the synthesis of cyclopropanes see: (a) Wendisch, D. In "Methoden der Organischen Chemie (Houben-Weyl)", 4th ed.; Mueller, E., Ed.; Georg Thieme: Stuttgart, 1971; Vol. IV, Part 3, pp 1-673; (b) Simmons, H. E.; Cairns, T. L.; Vladuchick, S. A.; Hoiness, C. M. *Org. React.* **1973**, *20,* 1; (c)

Boyle, P. H. In "Rodd's Chemistry of Carbon Compounds", 2nd ed.; Ansell, M. F., Ed.; Elsevier: Amsterdam, 1974; Vol. IIA, Suppl, pp 9-47 and the earlier parts of this series; (d) Yanovskaya, L. A.; Dombrovskii, V. A. *Russ. Chem. Rev.* **1975**, *44,* 154; (e) Moss, R. A. *Acc. Chem. Res.* **1980**, *13,* 58; (f) Freidlina, R. Kh.; Kamyshova, A. A.; Chukovskaya, E. Ts. *Russ. Chem. Rev.* **1982**, *51,* 368; (g) McQuillin, F. J.; Baird, M. S. "Alicyclic Chemistry", 2nd ed.; Cambridge University Press: Cambridge, 1983; pp 92-112; (h) Moody, C. J. In "Organic Reaction Mechanisms 1982"; Knipe, A. C.; Watts, W. E., Eds.; Wiley: Chichester, 1984; Chapter 6 and the earlier volumes of this series; (i) Hudlicky, T.; Kutchan, T. M.; Naqvi, S. M. *Org. React.* **1985**, *33,* 247; (j) Larock, R. C. "Organomercury Compounds in Organic Synthesis"; Springer-Verlag: Berlin, 1985; pp 327-380; (k) Doyle, M. P. *Chem. Rev.* **1986**, *86,* 919; (l) Brookhart, M.; Studabaker, W. B. *Chem. Rev.* **1987**, *87,* 411.

10. For some reviews of carbene complexes see: (a) Brown, F. J. *Prog. Inorg. Chem.* **1980**, *27,* 1; (b) Dötz, K. H.; Fischer, H.; Hofmann, P.; Kreissl, F. R.; Schubert, U.; Weiss, K. "Transition Metal Carbene Complexes"; Verlag Chemie: Weinheim, 1983; (c) Hahn, J. E. *Prog. Inorg. Chem.* **1984**, *31,* 205; (d) Casey, C. P. In "Reactive Intermediates"; Jones, M.; Moss, R. A., Eds.; Wiley: New York, 1985; Vol. 3 and the earlier volumes of this series; (e) Pourreau, D. B.; Geoffroy, G. L. *Adv. Organomet. Chem.* **1985**, *24,* 249; (f) Smith, A. K. *Organometal. Chem.* **1986**, *14,* 278 and the earlier volumes of this series. See also refs. 9k-m.

11. For some examples of carbene complexes undergoing cyclopropanation reactions with alkenes see: (a) Tebbe, F. N.; Parshall, G. W.; Reddy, G. S. *J. Am. Chem. Soc.* **1978**, *100,* 3611; (b) Casey, C. P.; Polichnowski, S. W.; Shusterman, A. J.; Jones, C. R. *J. Am. Chem. Soc.* **1979**, *101,* 7282; (c) Dötz, K. H.; Pruskil, I. *Chem. Ber.* **1981**, *114,* 1980; (d) Weiss, K.; Hoffmann, K.; *J.*

Organomet. Chem. **1983**, *255,* C24; (e) Toledano, C. A.; Rudler, H.; Daran, J.-C.; Jeannin, Y. *J. Chem. Soc., Chem. Commun.* **1984**, 574; (f) Casey, C. P.; Vollendorf, N. W.; Haller, K. J. *J. Am. Chem. Soc.* **1984**, *106,* 3754; (g) Doyle, M. P.; Griffin, J. H.; Bagheri, V.; Dorow, R. L. *Organometallics* **1984**, *3,* 53; (h) Mackenzie, P. B.; Ott, K. C.; Grubbs, R. H. *Pure Appl. Chem.* **1984**, *56,* 59; (i) Jacobson, D. B.; Freiser, B. S. *J. Am. Chem. Soc.* **1985**, *107,* 2605; (j) Doyle, M. P.; Griffin, J. H.; da Conceicao, J. *J. Chem. Soc., Chem. Commun.* **1985**, 328; (k) Doyle, M. P.; Dorow, R. L.; Terpstra, J. W.; Rodenhouse, R. A. *J. Org. Chem.* **1985**, *50,* 1663; (l) Parlier, A.; Rudler, H.; Platzer, N.; Fontanille, M.; Soum, A. *J. Organometal. Chem.* **1985**, *287,* C8; (m) Casey, C. P.; Shusterman, A. J. *Organometallics* **1985**, *4,* 736. See also the reviews cited in ref. 10.

12. (a) Jolly, P. W.; Pettit, R. *J. Am. Chem. Soc.* **1966**, *88,* 5044; (b) Riley, P. E.; Capshew, C. E.; Pettit, R.; Davis, R. E. *Inorg. Chem.* **1978**, *17*, 408.

13. Greene, M. L. H.; Ishaq, M.; Whiteley, R. N. *J. Chem. Soc. A* **1967**, 1508.

14. (a) Davison, A.; Krusell, W. C.; Michaelson, R. C. *J. Organomet. Chem.* **1974**, *72,* C7; (b) Sanders, A.; Bauch, T.; Magatti, C. V.; Lorenc, C.; Giering, W. P. *J. Organomet. Chem.* **1976**, *107,* 359; (c) Stevens, A. E.; Beauchamp, J. L. *J. Am. Chem. Soc.* **1978**, *100,* 2584; (d) Schilling, B. E. R.; Hoffmann, R.; Lichtenberger, D. L. *J. Am. Chem. Soc.* **1979**, *101,* 585; (e) Adams, R. D.; Davison, A.; Selegue, J. P. *J. Am. Chem. Soc.* **1979**, *101,* 7232; (f) Riley, P. E.; Davis, R. E.; Allison, N. T.; Jones, W. M. *J. Am. Chem. Soc.* **1980**, *102*, 2458; (g) Marten, D. F. *J. Chem. Soc., Chem. Commun.* **1980**, 341; (h) Brunner, H.; Leblanc, J.-C. *Z. Naturforsch. B: Anorg. Chem., Org. Chem.* **1980**, *35B,* 1491; (i) McCormick, F. B.; Angelici, R. J. *Inorg. Chem.* **1981**, *20,* 1111; (j) Bates, D. J.; Rosenblum, M.; Samuels, S. B. *J. Organomet. Chem.* **1981**, *209,* C55; (k) Manganiello, F. J.; Radcliffe, M. D.; Jones, W. M. *J. Organomet. Chem.* **1982**, *228*, 273; (l) Boland-Lussier, B. E.; Hughes, R. P. *Organometallics* **1982**, *1,* 635;

(m) Bodnar, T. W.; Cutler, A. R. *J. Am. Chem. Soc.* **1983**, *105,* 5926; (n) Curtis, P. J.; Davies, S. G. *J. Chem. Soc., Chem. Commun.* **1984**, 747; (o) Richmond, T. G.; Shriver, D. F. *Organometallics* **1984**, *3,* 305; (p) Bly, R. S.; Hossain, M. M.; Lebioda, L. *J. Am. Chem. Soc.* **1985**, *107,* 5549; (q) Guerchais, V.; Astruc, D. *J. Chem. Soc., Chem. Commun.* **1985**, 835; (r) Bodnar, T. W.; Cutler, A. R. *Organometallics* **1985**, *4,* 1558; (s) Barefield, E. K.; McCarten, P.; Hillhouse, M. C. *Organometallics* **1985**, *4,* 1682.

15. For leading references to work from the Brookhart laboratory see: (a) Brookhart, M.; Tucker, J. R.; Husk, G. R. *J. Am. Chem. Soc.* **1983**, *105,* 258; (b) Brookhart, M.; Timmers, D.; Tucker, J. R.; Williams, G. D.; Husk, G. R.; Brunner, H.; Hammer, B. *J. Am. Chem. Soc.* **1983**, *105,* 6721; (c) Brookhart, M.; Kegley, S. E.; Husk, G. R. *Organometallics* **1984**, *3,* 650; (d) Brookhart, M.; Studabaker, W. B.; Husk, G. R. *Organometallics* **1985**, *4,* 943.

16. (a) Casey, C. P.; Miles, W. H.; Tukada, H.; O'Connor, J. M. *J. Am. Chem. Soc.* **1982**, *104,* 3761; (b) Casey, C. P.; Miles, W. H. *Organometallics* **1984**, *3,* 808; (c) Casey, C. P.; Miles, W. H.; Tukada, H. *J. Am. Chem. Soc.* **1985**, *107,* 2924.

17. (a) Brandt, S.; Helquist, P. *J. Am. Chem. Soc.* **1979**, *101,* 6473; (b) O'Connor, E. J.; Helquist, P. *J. Am. Chem. Soc.* **1982**, *104,* 1869; (c) Kremer, K. A. M.; Helquist, P.; Kerber, R. C. *J. Am. Chem. Soc.* **1981**, *103,* 1862; (d) Kremer, K. A. M.; Helquist, P. *J. Organomet. Chem.* **1985**, *285,* 231.

18. (a) Kremer, K. A. M.; Kuo, G.-H.; O'Connor, E. J. Helquist, P.; Kerber, R. C. *J. Am. Chem. Soc.* **1982**, *104,* 6119; (b) Kuo, G.-H.; Helquist, P.; Kerber, R. C. *Organometallics* **1984**, *3,* 806.

19. Simmons, H. E.; Smith, R. D. *J. Am. Chem. Soc.* **1959**, *81,* 4256.

20. (a) Roth, J. A. *J. Catal.* **1972**, *26,* 97; (b) Kawabata, N.; Kamemura, I.; Naka, M. *J. Am. Chem. Soc.* **1979**, *101,* 2139.

21. (a) Corey, E. J.; Chaykovsky, M. *J. Am. Chem. Soc.* **1965**, *87*, 1353; (b) Becker, R. S.; Edwards, L.; Bost, R.; Elam, M.; Griffin, G. *J. Am. Chem. Soc.* **1972**, *94*, 6584.
22. (a) Nefedov, O. M.; Shafran, R. N.; Novitskaya, N. N. *J. Org. Chem. USSR* **1972**, *8*, 2121; (b) Vydrina, T. K.; Dolgoplosk, B. A.; Dolgii, I. E.; Nefedov, O. M.; Tinyakova, E. I. *Bull. Acad. Sci. USSR, Div. Chem.* **1977**, 217.
23. Oshima, K.; Shirafuji, T.; Yamamoto, H.; Nozaki, H. *Bull. Chem. Soc. Japan* **1973**, *46*, 1233.
24. (a) Bumgardner, C. L. *J. Am. Chem. Soc.* **1961**, *83*, 4420; (b) Martini, Th.; Kampmeier, J. A. *Angew. Chem., Int. Ed. Engl.* **1970**, *9*, 236.
25. Murayama, E.; Kikuchi, T.; Sasaki, K.; Sootome, N.; Sato, T. *Chem. Lett.* **1984**, 1897.
26. Jorgenson, M. J.; Thacher, A. F. In "Organic Syntheses. Collective Volume V"; Baumgarten, H. E., Ed.; Wiley: New York, 1973; pp 509-513.
27. Cerny, M.; Malek, J. *Coll. Czech. Chem. Commun.* **1976**, *41*, 119.
28. (a) Leonova, T. V.; Shapiro, I. O.; Ranneva, Yu. I.; Shatenshtein, A. I.; Shabarov, Yu. S. *J. Org. Chem. USSR* **1977**, *13*, 491; (b) Levina, R. Ya.; Shabarov, Yu. S.; Shanazarov, K. S.; Treshchova, E. G. *Vestn. Moskow Univ., Ser. Mat., Mekh., Astron., Fiz. i Khim.* **1957**, *12*, 145.
29. Miller, R. D.; Gölitz, P.; Janssen, J.; Lemmens, J. *J. Am. Chem. Soc.* **1984**, *106*, 7277.
30. Wieland, H.; Probst, O. *Justus Liebigs Ann. Chem.* **1937**, *530*, 274.
31. Wender, P. A.; Eck, S. L. *Tetrahedron Lett.* **1982**, *23*, 1871.

TABLE

OTHER CYCLOPROPANATIONS USING $C_5H_5(CO)_2FeCH_2S^+(CH_3)_2$ BF_4^-

Alkene	Cyclopropane Product(s)	Consumption of Alkene (%)[a]	Yield (%)[b]
		96	92
		81	70
		100	58
Ph	Ph	86	96
CO_2CH_3	CO_2CH_3 + CO_2CH_3	99	62 + 5
CO_2CH_3	CO_2CH_3	100	86[c]

[a] The % consumption values were determined by quantitative GLPC measurement of unreacted alkenes using an internal standard. [b] The yields were determined by quantitative GLPC using an internal standard and are corrected for unreacted alkenes. The reactions were typically run on 1-mmol scales. [c] Taken from ref. 31.

Acknowledgment. We wish to express our appreciation to the National Science Foundation, the National Institutes of Health, the Petroleum Research Fund administered by the American Chemical Society, the State University of New York at Stony Brook, and the University of Notre Dame for providing generous financial support for this work.

Appendix
Chemical Abstracts Nomenclature (Collective Index Number); (Registry Number)

1,1-Diphenylcyclopropane: Cyclopropane, 1,1-diphenyl- (8); Benzene, 1,1'-cyclopropylidenebis- (9); (3282-18-6)

Cyclopentadienyliron dicarbonyl dimer: Iron, tetracarbonylbis (η^5-2,4-cyclopentadien-1-yl)di-, (Fe-Fe) (9); (38117-54-3)

Sodium (8,9); (7440-23-5)

Chloromethyl methyl sulfide: Sulfide, chloromethyl methyl (8); methane, chloro(methylthio)- (9); (2373-51-5)

Iodomethane: Methane, iodo- (8,9); (74-88-4)

Sodium tetrafluoroborate: Borate (1-), tetrafluoro-, sodium (8,9); (13755-29-8)

Cyclopentadienyliron dicarbonyl dimethsulfonium tetrafluoroborate [i.e.(η^5-C$_5$H$_5$)(CO)$_2$FeCH$_2$S$^+$(CH$_3$)$_2$BF$_4^-$]: Iron (1+), dicarbonyl(η^5-2,4-cyclopentadien-1-yl)(dimethylsulfonium η-methylide)-, tetrafluoroborate (1-) (10); (72120-26-4)

1,1-Diphenylethylene: Ethylene, 1,1-diphenyl- (8); Benzene, 1,1'-ethylidenebis- (9); (530-48-3)

Dioxane: p-Dioxane (8); 1,4-Dioxane (9); (123-91-1)

Nitromethane: Methane, nitro- (8,9); (75-52-5)

1,2-ADDITION OF A FUNCTIONALIZED ZINC-COPPER ORGANOMETALLIC [RCu(CN)ZnI] TO AN α,β-UNSATURATED ALDEHYDE: (E)-2-(4-HYDROXY-6-PHENYL-5-HEXENYL)-1H-ISOINDOLE-1,3(2H)-DIONE

(1H-Isoindole-1,3(2H)-dione, 2-(4-hydroxy-6-phenyl-5-hexenyl)-)

A. 1 + NaI / acetone → 2

B. 2 + 1) Zn/THF; 2) CuCN·2 LiCl → 3

C. 3 + 1) BF$_3$·OEt$_2$; 2) PhCH=CHCHO → 4

Submitted by Ming Chang P. Yeh, Huai Gu Chen, and Paul Knochel.[1]
Checked by Thomas Wagler, Brian E. Jones, Thomas C. Zebovitz, and David L. Coffen.

1. Procedure

A. 2-(3-Iodopropyl)-1H-isoindole-1,3(2H)-dione **2**. A dry, one-necked, 500-mL, round-bottomed flask equipped with a magnetic stirring bar and a reflux condenser with a gas inlet at the top, is charged with 13.4 g (50 mmol) of 2-(3-bromopropyl)-1H-isoindole-1,3(2H)-dione **1** (Note 1), 17.95 g (120 mmol) of sodium iodide (Note 2), and 100 mL of acetone. The reaction mixture is stirred at reflux under nitrogen for 21 hr (Note 3). The solvent is removed on a rotary evaporator and the resulting solid is dissolved in 300 mL of dichloromethane and 200 mL of water. The two layers are separated in a separatory funnel and the aqueous layer is extracted with two 100-mL portions of dichloromethane. The combined organic extracts are washed successively with 100 mL of an aqueous 10% solution of sodium thiosulfate, three 100-mL portions of water and 150 mL of brine. The organic layer is dried over anhydrous magnesium sulfate. After filtration, the solvent is removed on a rotary evaporator. The crude white solid is dried for several hours at room temperature under reduced pressure to remove traces of solvent [15.0-15.5 g (47.6-49.2 mmol) 95-98% yield]. This material can be used directly in the next step (Note 4).

B. Formation of the copper-zinc organometallic **3** *from 2-(3-iodopropyl)-1H-isoindole-1,3(2H)-dione*. A dry, 100-mL, three-necked, round-bottomed flask is equipped with a magnetic stirring bar, 50-mL pressure equalizing addition funnel bearing a rubber septum, three-way stopcock, and a thermometer. The air in the flask is replaced by dry argon and the flask is charged with 4.71 g (72 mmol) of cut zinc (ca. 1.5 x 1.5 mm; Note 5). The flask is again flushed three times with argon. 1,2-Dibromoethane (Note 6), (0.2 mL, 2.3 mmol) and 3 mL of tetrahydrofuran (THF) (Note 7) are successively injected into the flask which is then heated gently with a heat gun until ebullition of solvent is observed; the zinc suspension is stirred a few minutes and heated again. The process is repeated three times; 0.15 mL (1.2 mmol) of chlorotrimethylsilane[2] is then injected into the addition funnel. The cut zinc foil turns grey. After 15 min the reaction mixture is heated to 30°C with an oil bath and 18.9 g (60 mmol) of iodide **2** dissolved in 30 mL of THF is added dropwise over 40 min. After addition, the reaction mixture is stirred for 4 hr at 43°C to give a dark brown-yellow solution of the zinc reagent (Note 8). A second, dry, 250-mL, three-necked, round-bottomed flask is equipped with a magnetic stirring bar, three-way stopcock connected to vacuum and an argon source, and two glass stoppers. The flask is charged with 4.59 g (108 mmol) of lithium chloride (Note 9). The flask is heated with an oil bath at 130°C (oil bath temperature) under vacuum (0.1 mm) for 2 hr to dry the lithium chloride. The reaction flask is then cooled to 25°C and flushed with argon. The two glass stoppers are replaced by a low temperature thermometer and a rubber septum and 4.84 g (54 mmol) of copper cyanide (Note 10) is added. The flask is flushed three times with argon and 40 mL of freshly-distilled THF (Note 7) is added to give, after 15 min, a clear yellow-green solution of the complex CuCN·2LiCl (Note 11). This solution is cooled to ca. -40°C and the two flasks are connected via a stainless steel cannula. The solution of the zinc reagent is transferred to the THF solution of copper cyanide and lithium chloride (Note 12). The resulting dark green solution is warmed to 0°C within 5 min and is ready to use in the next step after 5 min of stirring at this temperature.

C. *(E)-2-(4-Hydroxy-6-phenyl-5-hexenyl)-1H-isoindole-1,3(2H)-dione* **4**. The THF solution of the copper-zinc reagent is cooled to -78°C and 19.9 mL (162 mmol) of boron trifluoride etherate (Note 13) is added dropwise. The reaction mixture is warmed to -30°C and stirred for 30 min, then cooled to -60°C. (E)-Cinnamaldehyde (5.71 g, 43.2 mmol) is added slowly via a syringe. The reaction mixture is allowed to stir at -30°C for 14 hr (Note 14) and for 30 min at 0°C. After this time, conversion is complete as indicated by GLC analysis and the reaction mixture is poured into an Erlenmeyer flask containing 500 mL of ethyl acetate, 100 mL of a saturated aqueous solution of ammonium chloride and 5 mL of ammonium hydroxide. The mixture is filtered by suction through 10 g of Celite on a sintered glass funnel, the contents of the funnel are washed twice with 50 mL of ethyl acetate and the filtrate is separated into two layers. The organic layer is washed successively with 100 mL of aqueous 10% sodium thiosulfate, and twice with 100 mL of a saturated aqueous solution of ammonium chloride. The combined aqueous phases are extracted with 100 mL of ethyl acetate and the combined organic phases are washed with 100 mL of a saturated aqueous sodium chloride solution, then dried over magnesium sulfate. After filtration, the solvent is removed on a rotary evaporator (ca. 10 mm) to afford 19.85 g of a crude yellowish oil. Flash chromatography separation[4] of the oil using silica gel (230-400 mesh, 570 g) and ethyl acetate/hexane (1:2) gives 6.92 g (50% yield) of the 1,2-addition product **4** as a pale yellow solid, mp 87-88°C, after removal of the solvents (Note 15).

2. Notes

1. The N-(3-bromopropyl)phthalimide **1** was purchased from Aldrich Chemical Company, Inc. or from Lancaster Synthesis Ltd.

2. Sodium iodide (Analytical Reagent) was purchased from Mallinckrodt, Inc.

3. A GLC analysis (Megabore Column (DB5)) shows a conversion of 95%. The remaining bromide **1** is converted to the iodide **2** during the formation of the zinc organometallic (next reaction step).

4. The iodide **2** can be recrystallized from hexane/dichloromethane to give white needles; mp 87-88°C.[3] The spectra are as follows: IR (CH_2Cl_2) cm^{-1}: 3054.6 (m), 2892.5 (w), 1773.8 (m), 1716.1 (s), 1435.8 (s), 1396.3 (m), 1265.6 (s); ^{1}H NMR ($CDCl_3$, 360 MHz) δ: 2.25 (m, 2 H), 3.16 (t, 2 H, J = 7.2), 3.78 (t, 2 H, J = 3.6), 7.34 (dd,

2 H, J = 6.0 and 3.1); 7.86 (dd, 2 H, J = 6.0 and 3.1); ^{13}C NMR (CDCl$_3$, 90.5 MHz) δ: 2.1, 32.6, 38.2, 132.7, 134.8, 168.4.

5. This procedure uses cut zinc foil purchased from Alfa Products, Morton/Thiokol Inc. (foil, 0.25 mm thick, 30 cm wide, 99.9% purity). However, zinc dust can also be used. The reaction time using zinc dust is shorter and a lower reaction temperature may be possible. After formation of the zinc organometallic, the zinc dust is allowed to settle and the THF solution of the zinc organometallic is transferred via a syringe to the THF-solution of the complex CuCN·2LiCl. The checkers obtained a 51-56% yield of product, melting at 80-90°C.

6. 1,2-Dibromoethane and the chlorotrimethylsilane are purchased from Aldrich Chemical Company, Inc.

7. All the tetrahydrofuran used in this procedure was freshly distilled over sodium/benzophenone before use.

8. A GLC analysis of hydrolyzed aliquots allows one to check the completion of the reaction. Less than 7% of the starting iodide **2** and more than 93% of N-propylphthalimide can be detected. A yield of 90% of the zinc reagent is assumed. The zinc reagent has also been formed at a reaction temperature of 33°C. A reaction time of 16 hr is then required.

9. Anhydrous lithium chloride is purchased from Aldrich Chemical Company, Inc.

10. Copper cyanide, purchased from Aldrich Chemical Company, Inc., is not a hygroscopic salt and does not need to be dried before use.

11. A very small amount of undissolved lithium chloride may still be present and will dissolve after the addition of the zinc reagent.

12. To effect the transfer, the argon pressure in the flask containing the copper salt is reduced by inserting a needle through the septum and by shutting off the argon

gas entry. Washing the remaining zinc foil with 5 mL of dry THF allows one to transfer the zinc reagent almost quantitatively.

13. Boron trifluoride etherate is purchased from Aldrich Chemical Company, Inc., and is manipulated under argon.

14. An immersion cooler (Cryocool/Neslab) is used to maintain the temperature at -30°C.

15. A portion of this product is crystallized from 1:1 ethyl acetate:hexane to yield analytically pure product, mp 92-93°C (Anal. Calcd for $C_{20}H_{19}NO_3$: C, 74.75; H, 5.96; N, 4.36. Found: C, 74.52; H, 6.03; N, 4.29). The spectra are as follows: ^1H NMR (CDCl$_3$, 300 MHz) δ: 1.67-1.81 (m, 4 H), 3.75 (t, 2 H, J = 6.9), 4.34 (m, 1 H), 6.19 (dd, 1 H, J = 15.9 and 6.8), 6.56 (d, 1 H, J = 15.9), 7.22-7.36 (m, 5 H), 7.70 (m, 2 H), 7.82 (m, 2 H); ^{13}C NMR (CDCl$_3$, 75.5 MHz) δ: 24.6, 34.1, 37.7, 72.4, 123.1, 126.4, 127.6, 128.4, 128.5, 130.5, 131.9, 132.0, 133.8, 168.3; IR (CH$_2$Cl$_2$) cm^{-1}: 3489.5 (br), 3058.2 (m), 3024.8 (m), 2942.8 (m), 1771.3 (s), 1709.9 (s), 1467.6 (m), 1438.8 (s), 1398.8 (s), 1337.1 (s), 1266.2 (s), 1069.3 (m), 969.0 (m). Mass spectra (E.I.) m/e 321 (M$^+$, 45), 304 (5), 263 (2), 216 (24), 189 (20), 174 (87), 160 (89), 156 (25), 148 (33), 133 (100), 115 (30), 105 (41), 91 (44), 77 (45); High resolution M.S. Anal. Calcd for $C_{20}H_{19}NO_3$: 321.1365. Found: 321.1369.

Waste Disposal Information

All toxic materials were disposed of in accordance with "Prudent Practices for Disposal of Chemicals from Laboratories"; National Academy Press; Washington, DC, 1983.

3. Discussion

Organometallic compounds are among the most versatile intermediates for the formation of carbon-carbon bonds, but their high reactivity allows preparation of only relatively unfunctionalized reagents. In contrast, organozinc halides with a less reactive carbon-metal bond, display a high functional group tolerance, and can include a variety of functional groups such as esters,[5,6] enoates,[5b,7] ketones,[5a,8,9] nitriles,[10] halides,[11] amino groups,[7,12] phosphonates,[13] thioethers,[14] sulfoxides,[14] and sulfones.[14] A transmetallation of these zinc organometallics to the corresponding copper compounds, carried out using the THF-soluble copper salt CuCN·2 LiCl, affords highly reactive copper reagents [RCu(CN)ZnX]. In this procedure, we describe the synthesis of an alkylzinc iodide with a phthalimido group at the γ-position, its conversion to the corresponding copper derivative, and its regiospecific 1,2-addition to cinnamaldehyde in the presence of boron trifluoride etherate. Copper reagent 3 reacts with several other electrophiles in excellent yields (see Scheme). This preparation illustrates the convenient synthesis of highly functionalized organozinc halides in THF[15] and their high synthetic potential.

Scheme

1. The Willard H. Dow Laboratories, Department of Chemistry, The University of Michigan, Ann Arbor, MI 48109.
2. For various activations of zinc metal see: Erdik, E. *Tetrahedron* **1987**, *43*, 2203; for zinc activation with chlorotrimethylsilane, see: (a) Gawronski, J. K. *Tetrahedron Lett.* **1984**, *25*, 2605; (b) Picotin, G.; Miginiac, P. *J. Org. Chem.* **1987**, *52*, 4796; (c) Picotin, G.; Miginiac, P. *Tetrahedron Lett.* **1987**, *28*, 4551.
3. Martinkus, K. J.; Tann, C.-H.; Gould, S. J. *Tetrahedron* **1983**, *39*, 3493.
4. Still, W. C.; Kahn, M.; Mitra, M. *J. Org. Chem.* **1978**, *43*, 2923.
5. (a) Knochel, P.; Yeh, M. C. P.; Berk, S. C.; Talbert, J. *J. Org. Chem.* **1988**, *53*, 2390; (b) Yeh, M. C. P.; Knochel, P.; Butler, W. M.; Berk, S. C. *Tetrahedron Lett.* **1988**, *29*, 6693; (c) Majid, T. N.; Knochel, P. *Tetrahedron Lett.* **1990**, *31*, 4413.
6. (a) Tamaru, Y.; Ochiai, H.; Nakamura, T.; Tsubaki, K.; Yoshida, Z.-i.*Tetrahedron Lett.* **1985**, *26*, 5559; (b) Tamaru, Y.; Ochiai, H.; Nakamura, T.; Yoshida, Z.-i. *Tetrahedron Lett.* **1986**, *27*, 955; (c) Nakamura, E.; Sekiya, K.; Kuwajima, I. *Tetrahedron Lett.* **1987**, *28*, 337; (d) Tamaru, Y.; Ochiai, H.; Nakamura, T.; Yoshida, Z.-i. *Org. Synth.* **1989**, *67*, 98.
7. Yeh, M. C. P.; Knochel, P.; Santa, L. E. *Tetrahedron Lett.* **1988**, *29*, 3887.
8. Berk, S. C.; Knochel, P.; Yeh, M. C. P. *J. Org Chem.* **1988**, *53*, 5789.
9. Tamaru, Y.; Ochiai, H.; Nakamura, T.; Yoshida, Z.-i. *Angew. Chem.* **1987**, *99*, 1193; *Angew. Chem., Intern. Ed. Engl.* **1987**, *26*, 1157.
10. (a) Yeh, M. C. P.; Knochel, P. *Tetrahedron Lett.* **1988**, *29*, 2395; (b) Yeh, M. C. P.; Knochel, P. *Tetrahedron Lett.* **1989**, *30*, 4799; (c) Tamaru, Y.; Tanigawa, H.; Yamamoto, T.; Yoshida, Z.-i. *Angew. Chem.* **1989**, *101*, 358; *Angew. Chem., Intern. Ed. Engl.* **1989**, *28*, 351; (d) Majid, T. N.; Yeh, M. C. P.; Knochel, P. *Tetrahedron Lett.* **1989**, *30* 5069.
11. (a) Comins, D. L.; O'Connor, S. *Tetrahedron Lett.* **1987**, *28*, 1843; (b) Comins, D. L.; Foley, M. A. *Tetrahedron Lett.* **1988**, *29*, 6711.

12. (a) Chen. H. G.; Hoechstetter, C.; Knochel, P. *Tetrahedron Lett.* **1989**, *30*, 4795; (b) Jackson, R. F. W.; James, K.; Wythes, M. J.; Wood, A. *J. Chem. Soc., Chem. Commun.* **1989**, 644.
13. Retherford, C.; Chou, T.-S.; Schelkun, R. M.; Knochel, P. *Tetrahedron Lett.* **1990**, *31*, 1833.
14. AchyuthaRao, S.; Tucker, C. E.; Knochel, P. submitted for publication.
15. Gaudemar, M. *Bull. Soc. Chim. France* **1962**, 974.

Appendix
Chemical Abstracts Nomenclature (Collective Index Number); (Registry Number)

(E)-2-(4-Hydroxy-6-phenyl-5-hexenyl)-1H-isoindole-1,3(2H)-dione: 1H-Isoindole-1,3(2H)-dione, 2-(4-hydroxy-6-phenyl-5-hexenyl)- (12); (121883-31-6)

2-(3-Iodopropyl)-1H-isoindole-1,3(2H)-dione: 1H-Isoindole-1,3(2H)-dione, 2-(3-iodopropyl)- (9); (5457-29-4)

2-(3-Bromopropyl)-1H-isoindole-1,3(2H)-dione: Phthalimide, N-(3-bromopropyl)- (8); 1H-Isoindole-1,3(2H)-dione, 2-(3-bromopropyl)- (9); (5460-29-7)

Sodium iodide (8,9); (7681-82-5)

Zinc (8,9); (7440-66-6)

1,2-Dibromoethane: Ethane, 1,2-dibromo- (8,9); (106-93-4)

Chlorotrimethylsilane: Silane, chlorotrimethyl- (8,9); (75-77-4)

Lithium chloride (8,9); (7447-41-8)

Copper cyanide (8,9); (544-92-3)

Boron trifluoride etherate: Ethyl ether, compd. with boron fluoride (BF_3) (1:1) (8); Ethane, 1,1'-oxybis-, compd. with trifluoroborane (1:1) (9); (109-63-7)

(E)-Cinnamaldehyde: Cinnamaldehyde, (E)- (8); 2-Propenal, 3-phenyl-, (E)- (9); (14371-10-9)

SPIROANNELATION VIA ORGANOBIS(CUPRATES): 9,9-DIMETHYLSPIRO[4.5]DECAN-7-ONE

(Spiro[4.5]decan-7-one, 9,9-dimethyl-)

Submitted by Paul A. Wender,[1,2] Alan W. White,[1] and Frank E. McDonald.[2]
Checked by Naoki Hirayama and Hisashi Yamamoto.

1. Procedure

Note: All reactions should be conducted in an efficient fume hood.

A. **3-Chloro-5,5-dimethylcyclohex-2-en-1-one (1)**[3] (Note 1). An oven-dried, 250-mL, one-necked, round-bottomed flask is equipped with a magnetic stirring bar and graduated addition funnel topped with a nitrogen inlet. The flask is charged with dimedone (28.1 g, 200 mmol) and toluene (100 mL) (Note 2). The suspension is stirred while oxalyl chloride (35 mL, 400 mmol) is slowly added via the addition funnel over a 10-min period (Note 3). After the addition is complete and gas evolution has subsided, the addition funnel is quickly exchanged for a reflux condenser topped with a nitrogen inlet. The mixture is then heated at 60-70°C for 30 min, or until no more suspended dimedone remains and gas evolution has ceased. (Additional oxalyl chloride may be added until dimedone has completely reacted.) The reaction is allowed to cool and concentrated by rotary evaporation at reduced pressure. The crude red oil is distilled through a short path apparatus to give 3-chloro-5,5-dimethylcyclohex-2-en-1-one (1) (29.3 g, 93% yield) as a colorless oil, bp 68-71°C (6.0 mm) (Note 4).

B. **1,4-Dilithiobutane (2).**[4a,5] (All transfers are conducted under dry nitrogen; reagents are introduced into reaction vessels through rubber septa using a cannula or syringe.) An oven-dried, 1-L, three-necked, round-bottomed flask is equipped with a large magnetic stirring bar and glass beads (ca. 3-mm diameter), graduated addition funnel, stopper, and large diameter nitrogen inlet (at least 2 mm in diameter). The flask is purged with nitrogen, charged with anhydrous diethyl ether (250 mL) (Note 2), and cooled to 0°C. The stopper is removed from the flask and replaced with a conical funnel while a rapid flow of dry nitrogen is passed through the flask. Lithium wire, 1% Na (9.48 g, 1.36 mol, 4.5 eq.) (Note 5), prewashed with hexanes, is held with forceps over the funnel and cut with clean scissors into pieces no larger than 2 mm in length

(Note 6) so that the freshly cut lithium pieces drop directly into the anhydrous ether. 1,4-Dichlorobutane (33.5 mL, 300 mmol) (Note 2) is then dissolved in anhydrous diethyl ether (85 mL) and introduced into the addition funnel; approximately 10% of this solution is introduced into the lithium/ether suspension, and the reaction is initiated by vigorous stirring. A white precipitate (LiCl) signaling initiation of the reaction should be apparent within 5 to 15 min, at which time the remainder of the solution is added dropwise over a 1 to 2-hr period (Note 7). The white suspension is rapidly stirred for 20 hr at 0°C.

The mixture is most conveniently filtered by gravity filtration through an oven-dried coarse (15 μM) sintered glass frit (Notes 8, 9). The concentration of 1,4-dilithiobutane (**2**) in ether is determined by titration with sec-butyl alcohol using 1,10-phenanthroline as indicator. The molarity of the solution obtained under these optimized conditions is approximately 1.7 M in "RLi", i.e., 0.9 M in 1,4-dilithiobutane (**2**) (Note 10). This solution is stable for several months when stored at -10°C under nitrogen.

C. *9,9-Dimethylspiro[4.5]decan-7-one* (**3**).[5] (All transfers are conducted under dry nitrogen; reagents are introduced into reaction vessels through rubber septa using a cannula or syringe.) An oven-dried, 2-L, three-necked, round-bottomed flask is equipped with a graduated addition funnel, overhead mechanical stirrer, and a nitrogen inlet. The flask is purged with nitrogen and charged with copper(I) thiophenoxide (36.7 g, 212 mmol) and anhydrous tetrahydrofuran (400 mL) (Note 2), and the heterogeneous suspension is mechanically stirred while cooling in a -78°C cold bath (dry ice-acetone). 1,4-Dilithiobutane (**2**), 0.87 (± 0.02) M in diethyl ether (122 mL, 106 mmol) is added via the addition funnel over 5 min, and then the reaction mixture is allowed to slowly warm to -15°C (Note 11) over a 20 to 45-min period, during which time the initial yellow color changes to brown-red with concomitant dissolution of copper thiophenoxide. The addition funnel is washed with a few milliliters of anhydrous tetrahydrofuran, and a solution of 3-chloro-5,5-dimethylcyclohex-2-en-1-one (**1**) (15.85 g, 100 mmol) in anhydrous tetrahydrofuran (250 mL) is added dropwise over a 1 to 2-hr period, while the temperature of the cold bath is maintained at -15°C to -20°C. The reaction turns olive-green and then black as the chloroenone is added. After the addition is complete, the cold bath is removed and the reaction flask is allowed to warm to room temperature.

After 30 to 45 min, the reaction mixture is opened to the air and poured into approximately 500 mL of saturated aqueous ammonium chloride solution, diluted with approximately 500 mL of diethyl ether washings, and allowed to stir for 10 to 15 min. The resulting mixture is filtered through a Büchner funnel, washing with small portions of diethyl ether (Note 12). The layers are separated in a separatory funnel, the aqueous layer is extracted with diethyl ether, and the combined organic layers are washed with water, saturated aqueous sodium bicarbonate, and saturated aqueous sodium chloride, dried over approximately 100 g of sodium sulfate, filtered through a Büchner funnel, and concentrated by rotary evaporation. The concentrated product may still contain solid diphenyl disulfide that can now be efficiently removed by chromatography of the neat crude product mixture through a 5-cm diameter x 10-cm height silica gel column and elution with hexane-diethyl ether (7:1) (Note 13). Evaporation of solvent by rotary evaporation at reduced pressure gives 13.28 g (74% yield) of 9,9-dimethylspiro[4.5]decan-7-one (**3**) as a pale yellow to colorless oil (Note 14).

2. Notes

1. This procedure is identical to that originally published by Heathcock and Clark,[3] except that toluene has been substituted for benzene and chloroform as the solvent, because of the relative health hazards associated with the latter two solvents.

2. Dimedone, oxalyl chloride, 1,4-dichlorobutane, and copper thiophenoxide were purchased from Fluka Chemical Corporation, and were used without further purification. The checkers purchased dimedone, oxalyl chloride and 1,4-dichlorobutane from Nacalai Tesque, Inc., Kyoto, Japan and Tokyo Kasei Kogyo Co., LTD, Japan, and prepared copper thiophenoxide from thiophenol and copper(I) oxide. Toluene, diethyl ether and tetrahydrofuran were distilled from sodium-benzophenone ketyl immediately prior to use.

3. The addition of oxalyl chloride was accompanied by much gas evolution, but no apparent exothermic reaction. Two equivalents of oxalyl chloride were required in order to consume completely the dimedone.[3]

4. The spectral properties of **1** were as follows: ^{1}H NMR (400 MHz, CDCl$_3$) δ: 1.10 (s, 6 H), 2.26 (s, 2 H), 2.57 (d, 2 H, J = 1.4), 6.23 (t, 1 H, J = 1.4); IR (film) cm^{-1}: 2980 (m), 1680 (s), 1616 (m), 1346 (m), 1300 (m), 1276 (m), 1008 (m). The submitters obtained 30.1 g (95% yield) of **1**, bp 79-80°C (7.5 mm).

5. Lithium wire was obtained from Aldrich Chemical Company, Inc. The use of 4.5 equiv of lithium represented a 12.5% excess. The use of only 4 equiv of lithium gave a lower titer of 1,4-dilithiobutane (**2**), and a small amount of unreacted lithium always remained even after prolonged reaction times.

6. The lithium wire must be freshly cut and in pieces not exceeding 2 mm in length. The yield dropped sharply when the average length of lithium wire was increased to 5 mm. The preparation of 1,4-dilithiobutane (**2**) from 1,4-dichlorobutane failed with the use of lithium shot[4a] or low-sodium (<0.8% Na) lithium wire.

7. We have not yet observed an exothermic reaction in the initiation of this reaction, although maintaining the temperature at 0°C might help to control safely the lithiation reaction as well as to maximize the yield of 1,4-dilithiobutane (**2**).

8. Gravity filtration was preferred over vacuum filtration, since the latter method tended to pull LiCl through the frit. Small amounts of LiCl did not interfere with the formation or reaction of the biscuprate generated in Section C. The checkers used this solution without filtration.

9. In order to quench the small amount of unreacted lithium wire remaining in the reaction flask, the stopper was replaced by a reflux condenser open to the atmosphere at the top. Approximately 100 mL of diethyl ether was added to the reaction flask containing the lithium and the flask was cooled to 0°C under a stream of nitrogen. A 4:1 mixture of t-butyl alcohol : water was then added dropwise via the addition funnel until all of the lithium wire was consumed. *Caution: The quench is exothermic and is accompanied by the evolution of large amounts of hydrogen gas.* The mixture was then transferred to a separatory funnel for separation of the organic and aqueous layers followed by disposal.

10. Significant amounts of ether solvent are lost presumably by evaporation during the nitrogen flush and/or filtration steps. Thus, the molarity of the 1,4-dilithiobutane (**2**) solution is not an accurate indication of yield. The submitters titrated with menthol instead of with sec-butyl alcohol

11. Temperature control of the cold bath at -15°C was accomplished by addition of small amounts of dry ice to acetone and monitoring with a low-temperature thermometer. A slurry of dry ice in ethylene glycol was occasionally used as a -15°C cold bath.

12. The omnipresent solid contaminant was diphenyl disulfide, which was sparingly soluble in diethyl ether. Each filtration noted in the text was necessary for a successful workup on this large scale. The submitters used a medium (90 μm) sintered glass frit for these filtrations. The attempted removal of product **3** by distillation from diphenyl disulfide was largely unsuccessful because of efficient entrainment of **3** in diphenyl disulfide.

13. Pure **3** is best obtained by chromatography. Product **3** could also be purified by vacuum distillation through a 10-cm Vigreux column, bp 100-103°C (2.2 mm). However, distillation did not efficiently separate **3** from diphenyl disulfide, and bumping was often a serious problem.

14. The spectral properties of **3** were as follows: ^{1}H NMR (400 MHz, CDCl$_3$) δ: 1.02 (s, 6 H), 1.42-1.68 (m, 8 H), 1.70 (s, 2 H), 2.18 (s, 2 H); IR (film) cm^{-1}: 2950 (s), 1710 (s), 1450 (m), 1370 (m), 1280 (m), 1230 (m). The submitters obtained 13.01 g (72% yield) of **3**.

Waste Disposal Information

All toxic materials were disposed of in accordance with "Prudent Practices for Disposal of Chemicals from Laboratories"; National Academy Press; Washington, DC, 1983.

3. Discussion

The procedure in Section C is representative of the synthesis of spirobicyclic systems featuring the reaction of bis(nucleophile) reagents with geminal bis(electrophile) acceptors. This strategy provides for formation of both carbon-carbon bonds of the new ring in a single step.

The starting material 3-chloro-5,5-dimethylcyclohex-2-en-1-one (**1**) is easily synthesized from dimedone by the general methodology developed by Clark and Heathcock.[3] The β–chlorine can also be replaced with a variety of carbon- and heteronucleophiles,[6] and β–chloroenones can be easily reduced by zinc/silver couple to the corresponding enone.[3b]

The formation of 1,4-dilithiobutane (**2**) was first described by West and Rochow.[4c] The original procedure was modified by Whitesides, et al., in their pioneering studies on the synthesis and reactivity of metallocyclopentanes.[4a-b] The methodology described in Section B is general for the synthesis of a variety of 1,4- and 1,5-dilithioalkanes, as evident in the Table below.[5]

The synthesis of 9,9-dimethylspiro[4.5]decan-7-one (**3**) uses the organobis(cuprate) derived from 1,4-dilithiobutane (**2**) as a bis(nucleophile) component, which is added to the bis(electrophile) 3-chloro-5,5-dimethylcyclohex-2-en-1-one (**1**).

This methodology provides for spiroannelation at a carbon beta to the ketone, and is a complementary protocol for the cyclization of α,ω–dihaloalkanes to the kinetic enolates of 1,3-cycloalkanedione enol ethers (at the alpha position).[7]

The methodology has been successfully extended with modifications to both the bis(nucleophile) and the bis(electrophile) components, as shown in the Table.[5]

1. Department of Chemistry, Harvard University, Cambridge, MA 02138.
2. Department of Chemistry, Stanford University, Stanford, CA 94305.
3. (a) Clark, R. D.; Heathcock, C. H. *Synthesis* **1974**, *47*; (b) Clark, R. D.; Heathcock, C. H. *J. Org. Chem.* **1976**, *41*, 636.
4. (a) McDermott, J. X.; White, J. F.; Whitesides, G. M. *J. Am. Chem. Soc.* **1976**, *98*, 6521; (b) McDermott, J. X.; Wilson, M. E.; Whitesides, G. M. *J. Am. Chem. Soc.* **1976**, *98*, 6529; (c) West, R.; Rochow, E. G. *J. Org. Chem.* **1953**, *18*, 1739.
5. (a) Wender, P. A.; White, A. W. *J. Am. Chem. Soc.* **1988**, *110*, 2218; (b) Wender, P. A.; Eck, S. L. *Tetrahedron Lett.* **1977**, 1245.
6. For example, see (a) Sato, N.; Ishiyama, T.; Miyaura, N.; Suzuki, A. *Bull. Soc. Chem. Jpn.* **1987**, *60*, 3471; Kienzle, F.; Minder, R. E. *Helv. Chim. Acta* **1987**, *70*, 1537.
7. Stork, G.; Danheiser, R. L.; Ganem, B. *J. Am. Chem. Soc.* **1973**, *95*, 3414.

TABLE I
SPIROANNELATION USING ORGANOBIS(CUPRATES)

Reagent M = CuSPh	Equiv.	Substrate	Product	Time hr	Yield %
M~~~M 4	1.1	1	3	1	(96)
4	1.1	(cyclohexenone-Cl) 5	(spiro ketone)	1	(85)
4	1.1 / 1.1	(methyl cyclohexenone-Cl)	(methyl spiro ketone)	1 / 18	(80) / (93)
4	1.1	(cyclopentenone-Br)	(spiro ketone)	1	(87)
4	1.1	(methyl cyclopentenone-Br)	(methyl spiro ketone)	1	(76)
M~~~~~M	4.0	5	(spiro[5.5])	1	(74)
M-CH2-C6H4-M (ortho)	1.2	5	(indane spiro)	1	(49)
M-CH2CH2-C6H4-M (ortho)	3.0	5	(tetralin spiro)	2	88
biphenyl-2,2'-M2	3.0 / 10.0	5	(fluorene spiro)	2 / 2	39 / 94
M-C(=CH2)-CH2CH2-M	1.1	5	(methylene spiro)	0.5	(66)
M-C(=CH2)-CH2CH2CH2-M	4.0	5	(methylene spiro[5.5])	0.5	(56)

Yields in parentheses were determined by internal standard gas chromatographic analysis.

Appendix

Chemical Abstracts Nomenclature (Collective Index Number); (Registry Number)

9,9-Dimethylspiro[4.5]decan-7-one: Sprio[4.5]decan-7-one, 9,9-dimethyl- (10); (63858-64-0)

3-Chloro-5,5-dimethylcyclohex-2-en-1-one: 2-Cyclohexen-1-one, 3-chloro-5,5-dimethyl- (8,9); (17530-69-7)

Dimedone: 1,3-Cyclohexanedione, 5,5-dimethyl- (8,9); (126-81-8)

Oxalyl chloride (8); Ethanedioyl dichloride (9); (79-37-8)

1,4-Dilithiobutane: Lithium, μ-tetramethylenedi- (8); μ-1,4-butanediyldi- (9); (2123-72-0)

Lithium (8,9); (7439-93-2)

1,4-Dichlorobutane: Butane, 1,4-dichloro- (8,9); (110-56-5)

sec-Butyl alcohol: 2-Butanol, (±)- (8,9); (15892-23-6)

1,10-Phenanthroline (8,9); (66-71-7)

Copper(1) thiophenoxide: Benzenethiol, copper(1+) salt (8,9); (1192-40-1)

ALKYNYL(PHENYL)IODONIUM TOSYLATES: PREPARATION AND STEREOSPECIFIC COUPLING WITH VINYLCOPPER REAGENTS. FORMATION OF CONJUGATED ENYNES. 1-HEXYNYL(PHENYL)IODONIUM TOSYLATE AND (E)-5-PHENYLDODEC-5-EN-7-YNE

(Iodine, 1-hexynyl(4-methylbenzenesulfonato-O)phenyl-) and (Benzene, (1-butyl-1-octen-3-ynyl)-, (E)-)

A. $C_4H_9C\equiv CH \xrightarrow[-70°C \to -20°C]{C_4H_9Li,\ THF} \xrightarrow[-70°C \to reflux]{(CH_3)_3SiCl} C_4H_9C\equiv CSi(CH_3)_3$

B. $C_4H_9C\equiv CSi(CH_3)_3\ +\ PhIO \xrightarrow[0°C \to 25°C]{BF_3\cdot Et_2O,\ CHCl_3} \xrightarrow[25°C]{aq.\ NaOTs}$

$C_4H_9C\equiv C\text{-}I^+\text{-}C_6H_5\quad p\text{-}CH_3C_6H_4SO_3^-$

C. $C_4H_9Br \xrightarrow[Et_2O]{Mg} \xrightarrow{CuBr\cdot SMe_2} \xrightarrow{C_6H_5C\equiv CH} \xrightarrow{C_4H_9C\equiv C\text{-}I^+\text{-}C_6H_5\ \ p\text{-}CH_3C_6H_4SO_3^-}$

$$\begin{array}{c} C_6H_5 \\ \diagdown \\ C_4H_9 \end{array} C=C \begin{array}{c} H \\ \diagup \\ C\equiv CC_4H_9 \end{array}$$

Submitted by Peter J. Stang and Tsugio Kitamura.[1]
Checked by Kazuaki Ishihara and Hisashi Yamamoto.

1. Procedure

A. 1-Trimethylsilyl-1-hexyne (Note 1). A dry, 500-mL, three-necked, round-bottomed flask (equipped with a magnetic stirrer system) is fitted with one 50-mL and one 125-mL pressure-equalizing addition funnel and a reflux condenser. The top of the condenser is mounted with a T-piece connected at one end to an argon outlet and at the other end to an oil bubbler. The apparatus is purged with dry argon and the reaction is carried out under an argon atmosphere. The flask is charged with 16.4 g (0.20 mol) of 1-hexyne (Note 2) and 200 mL of tetrahydrofuran (THF, Note 3), and the mixture is cooled to approximately -70°C with a dry ice-2-propanol slush bath. The 125-mL dropping funnel is filled with 80 mL (0.20 mol) of 1.63 M butyllithium in hexane (Note 4) which is added over a period of 30 min to the stirred, cold (-70°C) 1-hexyne in THF. After addition is complete the flask is gradually warmed over a 3-hr period to -20°C, then recooled to -70°C with the dry ice-2-propanol slush bath. Chlorotrimethylsilane (Note 5), 25.3 mL (0.20 mol), is placed in the 50-mL dropping funnel and added over a period of 15 min to the stirred, cooled reaction mixture. After the addition is complete the slush bath is removed and the mixture is first stirred at room temperature for 16 hr, then heated under reflux for 1 hr (Note 6). The mixture is cooled to 0°C with an ice-water bath and then 50 mL of water is carefully added. The entire reaction mixture is transferred to a 1-L separatory funnel and extracted with 200 mL of pentane. The organic phase is separated, washed successively with three 100-mL portions of water followed by 100 mL of saturated sodium chloride solution and then dried over anhydrous magnesium sulfate. After filtration the solvent is distilled off through a 15-cm Vigreux column. The pale yellow residue is transferred to a 100-mL round-bottomed flask and distilled through a 15-cm Vigreux column to give 25.0-27.5 g (81-89%) of 1-trimethylsilyl-1-hexyne as a clear, colorless liquid, bp 149-156°C, that can be used as is in the subsequent step.

B. *1-Hexynyl(phenyl)iodonium tosylate.* The center neck of a dry, 500-mL, three-necked, round-bottomed flask (equipped with a magnetic stirrer system) is fitted with a 50-mL pressure-equalizing addition funnel One side neck is fitted with a glass stopper and the other with a gas inlet to which is attached a T-piece connected to an argon supply and an oil bubbler. After the flask is purged with dry argon, it is charged with 22.0 g (0.10 mol) of finely ground iodosobenzene (Note 7), 200 mL of chloroform (Note 8) and 17.0 g (0.11 mol) of 1-trimethylsilyl-1-hexyne, and the entire mixture is cooled to 0°C with an ice-water bath. Boron trifluoride etherate (Note 9), 13.0 mL (0.11 mol), is added dropwise over a period of about 5 min to the stirred, cooled reaction mixture (Note 10). After the addition is complete the mixture is stirred at room temperature for about 16 hr (Note 11).

To a 1-L beaker are added 76.1 g (0.4 mol) of p-toluenesulfonic acid monohydrate (Note 12) and 400 mL of water. To this solution is carefully added 21.2 g (0.2 mol) of anhydrous sodium carbonate and then the entire solution is purged with argon for about 15 min to replace all air. This solution of sodium toluenesulfonate (NaOTs), along with the contents of the reaction flask, are placed in a 1-L separatory funnel and the mixture is vigorously shaken for a period of about 10 min (Note 13). The organic phase is separated and the aqueous phase is extracted with 50 mL of methylene chloride. The combined organic phases are dried over anhydrous magnesium sulfate, filtered and the solvent evaporated on a rotary evaporator. To the residual yellow-brown oil is added a mixture of 75 mL of diethyl ether and 75 mL of pentane resulting, after stirring, in fine white crystals. The powdery white crystals are filtered, washed with two 50-mL portions of a diethyl ether/pentane (1:1 v/v) mixture, air dried, then dried under reduced pressure. The resulting white crystalline solid, 27-34 g (60-75%), mp 77-80°C, (dec) is essentially pure and ready for most uses (Note 14).

C. *(E)-5-Phenyldodec-5-en-7-yne. Caution: Dimethyl sulfide is a volatile, very smelly irritant. This reaction must be conducted in a hood!*

A dry, 100-mL, three-necked, round-bottomed flask (equipped with a magnetic stirrer system) is fitted with a 25-cm Liebig condenser, a 25-mL pressure-equalizing dropping funnel topped off with a gas-inlet, and a glass stopper. After the flask is purged with argon, it is charged with 0.90 g (37.5 mmol) of magnesium turnings and 15 mL of anhydrous ether. To the dropping funnel is added a solution of 4.0 mL (37.5

mmol) of 1-bromobutane (Note 15) in 20 mL of anhydrous ether. This solution is added to the reaction flask at such a rate (approximately 1 hr) that after the Grignard reaction is initiated a gentle reflux is maintained. After the addition is complete the mixture is stirred for an additional 30 min.

The center neck of a dry, 250-mL, three-necked, round-bottomed flask (equipped with a magnetic stirrer system) is fitted with a 50-mL pressure-equalizing dropping funnel topped off with a rubber septum and an argon inlet. The side necks are fitted with a glass stopper and a rubber septum, respectively. The apparatus is purged with argon and to the flask are added 7.7 g (37.5 mmol) of copper(I) bromide-dimethyl sulfide complex (Note 16), 60 mL of anhydrous ether, and 38 mL of dimethyl sulfide (Note 17). The mixture is cooled to -70°C with a dry ice-2-propanol slush bath. The previously prepared butyl Grignard reagent is transferred to the addition funnel via a double-ended needle with a positive argon pressure. The Grignard reagent is added dropwise, over a 30-min period, to the cooled reaction mixture (Note 18). After the addition is complete the reaction is maintained between -40°C to -50°C for 2 hr, then recooled to -70°C. Phenylacetylene (Note 19), 4.1 mL (37.1 mmol), is slowly added via a syringe through the rubber septum. After addition, the reaction mixture is maintained at -30°C to -25°C for 2 hr (Note 20), then recooled to -70°C and 6.8 g (15 mmol) of the previously prepared iodonium tosylate is added, in the solid form, through the side neck, over a positive argon pressure. The stirring reaction mixture is gradually warmed to room temperature and stirred for an additional 12 hr at room temperature. The mixture is poured into a 1-L beaker containing 200 mL of saturated ammonium chloride solution and stirred vigorously. The undissolved materials are filtered off and the organic phase is separated. The aqueous phase is extracted with three 50-mL portions of ether. The combined organic phase are washed with 100 mL of water followed by 100 mL of saturated sodium chloride solution, and dried over anhydrous magnesium sulfate and filtered. The solvent is removed on a rotary evaporator in the hood and the residual oil chromatographed (4 x 50 cm column) on alumina (600 g, Note 21). The column is successively eluted with 1 L of hexane, 750 mL of 10% dichloromethane-hexane, 500 mL of 20% dichloromethane-hexane and 250 mL of 30% dichloromethane-hexane. The fractions are analyzed by TLC on silica gel (Note 22) using 10% dichloromethane-hexane as eluent and the product-containing fractions (Note 23) are combined. The solvent is evaporated on a rotary evaporator and the residual yellow-brown oil (5.8 g) (Note 24) is distilled under reduced pressure, bp 135-143°C (3-4 mm), to give 2.3 g of crude product. Redistillation under reduced pressure affords 1.9-2.1 g (53-58%) of product as a pale yellow liquid, bp 143-144°C (1 mm) (Note 25).

2. Notes

1. Many 1-trimethylsilylalkynes are commercially available from Aldrich, Petrarch and other vendors of silicon compounds. Alternatively they are readily made from 1-alkynes and chlorotrimethylsilane as described in part A.

2. 1-Hexyne was obtained from Tokyo Kasei Kogyo Co., Ltd. or Aldrich Chemical Company, Inc. and used without purification.

3. Tetrahydrofuran was obtained from Wako Pure Chemical Industries, Ltd., and distilled from potassium benzophenone ketyl immediately before use.

4. Butyllithium in hexane (2.6 M) was obtained from Mitsuwas Pure Chemical or Aldrich Chemical Company, Inc. and a freshly opened sample was used without assay or purification.

5. Chlorotrimethylsilane was obtained from Shin-ETSU Silicon Chemicals or PCR Research Chemicals, Inc., and distilled prior to use.

6. During this period copious amounts of a white precipitate (LiCl) is formed.

7. Iodosobenzene was purchased by the checkers from Tokyo Kasei Kogyo Co., Ltd. and dried under reduced pressure prior to use.

8. Chloroform was distilled from phosphorus oxide (P_2O_5) and passed through an alumina column prior to use.

9. Boron trifluoride etherate (98%) was obtained from Wako Pure Chemical Industries, Ltd., or MC and B Manufacturing Chemists, and distilled from granular calcium hydride prior to use.

10. The pale yellow suspension turns to a deeper yellow suspension during addition.

11. Iodosobenzene is a polymer and depolymerizes as it reacts, forming a clear, yellow-light brown, homogeneous solution during the course of the reaction. It is not ready for work-up until a clear homogeneous solution is obtained.

12. p-Toluenesulfonic acid monohydrate was obtained from Wako Pure Chemical Industries, Ltd., or MC and B Manufacturing Chemists, and used without further purification.

13. Care must be taken to release periodically the pressure formed. It is advisable to chill both solutions prior to mixing and shaking.

14. The spectral properties of 1-hexynyl(phenyl)iodonium tosylate are as follows: IR (nujol) cm^{-1}: 2185 (m, C≡C), 1225 (s), 1175 (m), 1145 (vs), 1115 (m), 1025 (m), 1000 (s), 985 (m, sh), 807 (m), 730 (m), 675 (s); FAB-MS m/z 285 (n, BuC≡CIPh$^+$); ^{1}H NMR (CDCl$_3$) δ: 0.73-0.92 (m, CH$_3$), 1.12-1.58 (m, CH$_2$CH$_2$), 2.30 (s, ArCH$_3$), 2.30-2.50 (m, CH$_2$). 6.98-7.60 (m, ArH), 7.92-8.05 (m, ArH).

15. 1-Bromobutane was obtained from Wako Pure Chemical Industries, Ltd., or Eastman Kodak Company, and distilled through a 10-cm Vigreux column prior to use.

16. Copper(I) bromide-dimethyl sulfide complex, (CuBr·SMe$_2$), 99%, was obtained from Aldrich Chemical Company, Inc. and purified prior to use as follows: 10 g of CuBr·SMe$_2$ was dissolved in 20 mL of dimethyl sulfide and triturated with 30 mL of

pentane. The resulting white crystals were filtered, washed with three 20-mL portions of pentane and dried under reduced pressure.

17. Dimethyl sulfide, 99+%, anhydrous, was obtained from Tokyo Kasei Kogyo Co., Ltd., or Aldrich Chemical Company, Inc.

18. The white suspension turned to an orange suspension during the course of the addition.

19. Phenylacetylene was obtained from Tokyo Kasei Kogyo Co., Ltd., or Farchan Laboratories and distilled through a 15-cm Vigreux column prior to use.

20. During this period the color changed from orange through brown to black.

21. Unactivated alumina from J. T. Baker Chemical Company was used.

22. TLC silica gel 60F$_{254}$ sheets were obtained from Merck & Company, Inc.

23. The desired product appears as the second spot on TLC with an R$_f$ = 0.42-0.45 with 10% dichloromethane-hexane.

24. The major by-product is 5,8-diphenyldodeca-5,7-diene, which can be separated by distillation.

25. GC analysis of the product on a 10% UCW-982 Chromosorb W column (0.25 in x 6 ft) at 200°C showed a single isomer with a purity of 99%. The spectral properties of the product are as follows: IR (neat) cm^{-1}: 3070 (m, sh), 3050 (m), 3015 (m), 2950 (vs), 2925 (vs), 2855 (s), 2200 (w), 1595 (m), 1570 (w), 1490 (m), 1465 (s), 1442 (s), 1375 (m), 1325 (m), 847 (m), 752 (vs), 690 (vs), MS (EI) m/z: 240 M+, 57%), 141 (100%), 105 (53%); ^{1}H NMR (CDCl$_3$) δ: 0.81-1.00 (m, CH$_3$), 1.21-1.59 (m, CH$_2$CH$_2$), 2.30-2.46 (m, CH$_2$), 2.67-2.84 (m, CH$_2$), 5.74 (t, J = 2, C=CH), 7.14-7.51 (m, ArH): ^{13}C NMR (300 MHz, CDCl$_3$) δ: 13.52 (CH$_3$), 13.81 (CH$_3$), 19.35 (CH$_2$), 21.91 (CH$_2$), 22.50 (CH$_2$), 30.50 (CH$_2$) 30.96 (CH$_2$), 31.70 (CH$_2$), 78.93 (C≡C), 95.47 (C≡C), 107.42 (C=$\underline{C}$H), 125.78, 127.44, 128.24, 140.52 (aromatic), 151.88 ($\underline{C}$=CH).

Waste Disposal Information

All toxic materials were disposed of in accordance with "Prudent Practices for Disposal of Chemicals from Laboratories"; National Academy Press; Washington, DC, 1983.

3. Discussion

The only known alternative procedure for the preparation of alkynyl(phenyl)iodonium arylsulfonates, the latest member of the family[2] of polyvalent iodine compounds, involves the reaction of [hydroxy(tosyloxy)iodo]benzene, PhIOH·OTs, with terminal alkynes as first reported by Koser[3] and elaborated by us.[4] This procedure has a number of shortcomings. Formation of the desired alkynyliodonium salt is usually accompanied by a related vinyl species, R(TsO)C=CHIPh·OTs, that both decreases the yields and causes purification problems. Furthermore, when the alkyl substituent of the starting alkyne is small, such as CH_3, n-Pr, n-Bu, etc., this procedure gives either no product[3] or low yields[4] at best.

The present procedure, similar to that of Fujita[5] for the preparation of alkynyl(phenyl)iodonium tetrafluoroborates, $RC{\equiv}CIPh{\cdot}BF_4$, is simpler, much more general and in most cases gives significantly better yields.[6] Table I gives yields of alkynyl(phenyl)iodonium sulfonates prepared by this procedure.

TABLE I

ALKYNYL(PHENYL)IODONIUM SULFONATES

Starting Alkyne	Product	Yields (%)	M.P. (dec.) °C
$CH_3C{\equiv}CH$	$CH_3C{\equiv}CIPh \cdot OTs$	62	123-125
$EtC{\equiv}CH$	$EtC{\equiv}CIPh \cdot OTs$	81	108-110
n-$PrC{\equiv}CH$	n-$PrC{\equiv}CIPh \cdot OTs$	89	93-95
$Me_3SiC{\equiv}CSiMe_3$	$Me_3SiC{\equiv}CIPh \cdot OTs$	70	107-109
$PhC{\equiv}CH$	$PhC{\equiv}CIPh \cdot OTs$	61	128-133[4]
t-$BuC{\equiv}CH$	t-$BuC{\equiv}CIPh \cdot OTs$	67	118-124[4]

Alkynyl(phenyl)iodonium sulfonates are stable, microcrystalline substances that can be stored and used indefinitely with little or no decomposition. They have been employed in the formation of aryl (2-furyl)iodonium tosylates,[7] alkynyl sulfonate,[4] carboxylate[8] and phosphate[9] esters, tricoordinate vinyliodinane species,[10] and alkylidenecarbene-iodonium ylides.[11]

The stereoselective formation of conjugated enynes has been reported[12] via the coupling of vinylcopper reagents with alkynyl(phenyl)iodonium tosylates. A representative example of this process is described in part C of the present procedure. This method[12] affords stereoisomerically pure 1,1-disubstituted conjugated 1,3-enynes with a trisubstituted olefin component with complete control of olefin geometry. The simplicity of the procedure, mild reaction conditions, reasonable yields (46-94%) and total stereocontrol[12] should make this an attractive alternative and complement to the known[13,14] Pd-catalyzed olefin-alkyne couplings for the synthesis of this important class of aliphatic compounds.

1. Department of Chemistry, The University of Utah, Salt Lake City, UT 84112. Financial support from the Cancer Institute of NIH (2ROCA16903) is gratefullly acknowledged.
2. Koser, G. F. In "The Chemistry of Halides, Pseudo-Halides and Azides" [The Chemistry of Functional Groups, Supplement D]; Patai, S.; Rappoport, Z., Eds.; Wiley: 1983, Part 2, Chapter 25, pp. 1265-1351.
3. Rebrovic, L.; Koser, G. F. *J. Org. Chem.* **1984**, *49*, 4700; Koser, G. F.; Rebrovic, L.; Wettach, R. H. *J. Org. Chem.* **1981**, *46*, 4324.
4. Stang, P. J.; Surber, B. W.; Chen, Z.-C.; Roberts, K. A.; Anderson, A. G. *J. Am. Chem. Soc.* **1987**, *109*, 228; Stang, P. J.; Surber, B. W. *J. Am. Chem. Soc.* **1985**, *107*, 1452.
5. Ochiai, M.; Kunishima, M.; Sumi, K.; Nagao, Y.; Fujita, E.; Arimoto, M.; Yamaguchi, H. *Tetrahedron Lett.* **1985**, *26*, 4501.
6. Kitamura, T.; Stang, P. J. *J. Org. Chem.* **1988**, *53*, 4105.
7. Margida, A. J.; Koser, G. F. *J. Org. Chem.* **1984**, *49*, 4703.
8. Stang, P. J.; Boehshar, M.; Wingert, H.; Kitamura, T. *J. Am. Chem. Soc.* **1988**, *110*, 3272.
9. Stang, P. J.; Boehshar, M.; Lin, J. *J. Am. Chem. Soc.* **1986**, *108*, 7832.
10. Stang, P. J.; Wingert, H.; Arif, A. M. *J. Am. Chem. Soc.* **1987**, *109*, 7235.
11. Kitamura, T.; Stang, P.J. *Tetrahedron Lett.* **1988**, *29*, 1887.
12. Stang, P. J.; Kitamura, T. *J. Am. Chem. Soc.* **1987**, *109*, 7561.
13. Stille, J. K.; Simpson, J. H. *J. Am. Chem. Soc.* **1987**, *109*, 2138.
14. Scott, W. J.; McMurry, J. E. *Acc. Chem. Res.* **1988**, *21*, 47; Heck, R. F. "Palladium Reagents in Organic Synthesis"; Academic Press: New York, 1985.

Appendix

Chemical Abstracts Nomenclature (Collective Index Number); (Registry Number)

1-Hexynyl(phenyl)iodonium tosylate: Iodine, 1-hexynyl(4-methylbenzenesulfonato-O)phenyl- (11); (94957-42-3)

(E)-5-Phenyldodec-5-en-7-yne: Benzene, (1-butyl-1-octen-3-ynyl)-, (E)- (12); (111525-79-2)

1-Trimethylsilyl-1-hexyne: Silane, 1-hexynyltrimethyl- (8,9); (3844-94-8)

1-Hexyne (8,9); (693-02-7

Chlorotrimethylsilane: Silane, Chlorotrimethyl- (8,9); (75-77-4)

Iodosobenzene: Benzene, iodoso- (8); Benzene, iodosyl- (9); (536-80-1)

Boron trifluoride etherate: Ethyl ether, compd. with boron fluoride (BF_3) (1:1) (8); Ethane, 1,1'-oxybis-, compd. with trifluoroborane (1:1) (9); (109-63-7)

p-Toluenesulfonic acid monohydrate (8); Benzenesulfonic acid, 4-methyl-, monohydrate (9); (6192-52-5)

1-Bromobutane: Butane, 1-bromo- (8,9); (109-65-9)

Dimethyl sulfide: Methyl sulfide (8); Methane, thiobis- (9); (75-18-3)

Phenylacetylene: Benzene, ethynyl- (8,9); (536-74-3)

2-METHYL-1,3-CYCLOPENTANEDIONE

(1,3-Cyclopentanedione, 2-methyl-)

Submitted by Philip G. Meister, Matthew R. Sivik, and Leo A. Paquette.[1]
Checked by David L. Coffen.

1. Procedure

Caution! These operations result in the evolution of considerable amounts of hydrogen chloride and should therefore be performed in a well-ventilated hood

A dry, 5-L, three-necked, round-bottomed flask equipped with a nitrogen inlet, mechanical stirrer (Note 1), and an efficient reflux condenser (Note 2) is charged with 500 mL of dry nitromethane (Note 3). Stirring is begun and 1000 g (7.50 mol) of anhydrous aluminum chloride (Note 4) is added, followed by an additional 500 mL of dry nitromethane. After the reaction mixture cools to room temperature, the gas inlet is replaced with a Gooch tube attached to a 500-mL filter flask containing 295 g (2.5 mol) of powdered succinic acid (Note 5). The nitrogen line is now attached to the sidearm of this flask. The succinic acid is introduced in portions during 1.5 hr (*Caution: This process evolves a large volume of hydrogen chloride gas which may cause the mixture to foam. Small quantities of the acid should be added at a time and the foaming should be allowed to subside prior to introduction of the next amount*). The mixture is stirred for 2 hr and the Gooch tube is replaced by a 500-mL pressure-equalizing addition funnel equipped with a nitrogen inlet. Propionyl chloride (650 mL,

694 g, 7.5 mol) (Note 6) is added dropwise during 30 min and the reaction mixture is brought to reflux for 2 hr, cooled, and poured onto 4 L of crushed ice. After the precipitated brown solid is cooled in an ice bath, it is separated by filtration (Note 7) and washed with 250 mL of brine and 250 mL of cold (0°C) toluene. The material is dissolved in 7 L of boiling water containing 20 g of decolorizing carbon, then filtered while still hot (Note 8). The filtrate is concentrated to a volume of 5 L, then cooled in an ice bath. The crystals are collected by suction filtration and air-dried to give 157-171 g (56-61%) of 2-methyl-1,3-cyclopentanedione (Note 9). The mother liquors are concentrated to approximately 1.5 L by rotary evaporation. The solution is boiled until crystals form, cooled in ice, and filtered to give an additional 20-23 g (7-8%) of product (63-69% overall yield).

2. Notes

1. Efficient stirring is mandatory.
2. The top of the condenser is equipped with a gas outlet leading via an oil bubbler into a large alkali bath or to a water aspirator. A steady stream of nitrogen should be maintained at all times.
3. Commercial nitromethane was dried over calcium chloride, filtered, and distilled. The forerun was discarded. The checker used 96% spectrophotometric grade as received from Aldrich Chemical Company, Inc.
4. Aluminum chloride, which was obtained from Fluka Chemical Corporation, generates heat during dissolution.
5. Succinic acid was obtained in powdered form from Fluka Chemical Corporation. If granular succinic acid is to be used, it should be pulverized to aid in dissolution.

6. Propionyl chloride was purchased from Fluka Chemical Corporation and Aldrich Chemical Company, Inc.

7. Two 17-cm Büchner funnels should be used simultaneously. However, if left to digest overnight, the mixture can be conveniently filtered using a 3-L sintered glass funnel.

8. Use of a 17-cm Büchner funnel preheated with hot tap water has proven most convenient for this purpose. The checker used an oven-heated, 3-L sintered glass funnel layered with Celite.

9. The crystals are off-white to tan, mp 211-212°C.

Waste Disposal Information

All toxic materials were disposed of in accordance with "Prudent Practices for Disposal of Chemicals from Laboratories"; National Academy Press; Washington, DC, 1983.

3. Discussion

A variety of methods have been published to achieve the preparation of 2-methyl-1,3-cyclopentanedione.[2-6] The present method, which was first reported by Schick, Lehmann, and Hilgetag,[4] provides for the acquisition of large amounts of the product in a single step from inexpensive starting materials. The alternative multistep procedures are appreciably more laborious and costly.

The importance of the title compound as an intermediate in organic synthesis goes unquestioned, having been produced on an industrial scale. The syntheses and reactions of this class of compounds have recently been summarized in an extensive review of the subject.[7]

The success of the present process appears to rest on the facility with which intermediate **1** is produced and its capacity for ready intramolecular cyclization.[4]

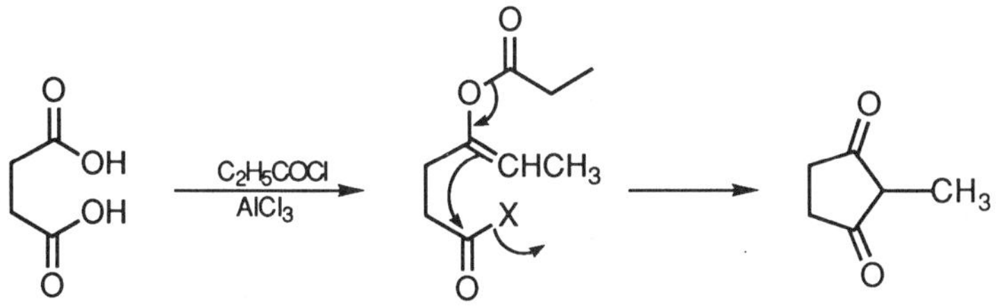

(X = OCOC$_2$H$_5$ or Cl)

1

1. Department of Chemistry, The Ohio State University, Columbus, OH 43210.
2. Orchin, M.; Butz, L. W. *J. Am. Chem. Soc.* **1943**, *65*, 2296.
3. Grenda, V. J.; Lindberg, G. W.; Wendler, N. L.; Pines, S. H. *J. Org. Chem.* **1967**, *32*, 1236.
4. Schick, H.; Lehmann, G.; Hilgetag, G. *Chem. Ber.* **1969**, *102*, 3238.
5. John, J. P.; Swaminathan, S.; Venkataramani, P. S. *Org. Synth., Coll. Vol. V* **1973**, 747.
6. Hengartner, U.; Chu, V. *Org. Synth., Coll. Vol. VI* **1988**, 774.
7. Schick, H.; Eichhorn, I. *Synthesis* **1989**, 477.

Appendix

Chemical Abstracts Nomenclature (Collective Index Number); (Registry Number)

2-Methyl-1,3-cyclopentanedione: 1,3-Cyclopentanedione, 2-methyl- (8,9); (765-69-5)

Aluminum chloride (8,9); (7446-70-0)

Nitromethane: Methane, nitro- (8,9); (75-52-5)

Succinic acid (8); Butanedioic acid (9); (110-15-6)

Propionyl chloride (8); Propanoyl chloride (9); (79-03-8)

PREPARATION OF (E,Z)-1-METHOXY-2-METHYL-3-(TRIMETHYLSILOXY)-1,3-PENTADIENE

(Silane, [[1-(2-methoxy-1-methylethenyl)-1-propenyl]oxy]trimethyl-, (Z,E)-)

A. 3-pentanone → (1. HCO₂Et, NaH, benzene; 2. H₃O⁺) → compound **1** (enol, OH)

B. compound 1 → (MeOH, TsOH, benzene) → compound **2** (OMe)

C. compound 2 → (TMSOTf, TEA, ether) → compound **3** (OTMS, OMe diene)

Submitted by David C. Myles and Mathew H. Bigham.[1,2]
Checked by David I. Magee and Robert K. Boeckman, Jr.

1. Procedure

Caution! Benzene has been identified as a carcinogen; OSHA has issued emergency standards on its use. All procedures involving benzene should be carried out in a well-ventilated hood, and glove protection is required.

A. A 5-L, three-necked, round-bottomed flask equipped with a mechanical stirrer, calcium sulfate drying tube, and 500-mL pressure equalizing addition funnel is charged with 2.4 L of dry benzene (Note 1), 81.7 g (2.04 mol) of 60% sodium hydride dispersion in benzene (Note 2), and 1.9 mL (0.05 mol) of methyl alcohol (Note 3). The flask is immersed in an ice bath and the addition funnel is charged with a mixture (Note 4) of 222 mL (172 g, 2 mol) of 95% 3-pentanone and 163 mL (148 g, 2 mol) of 99% ethyl formate (Note 5). This mixture is added dropwise to the cooled, stirred suspension over 1.5 hr. During the addition, there is a visible evolution of gas, the mixture darkens, and a paste-like precipitate forms. At the completion of the addition, the ice bath is removed and the mixture is stirred an additional 1.5 hr, until gas evolution has ceased. The stirred reaction mixture is then diluted with 1.5 L of anhydrous ether (Note 6). The suspension is filtered through an 18.5-cm Büchner funnel fitted with two Whatman 1 filter papers and the solid is washed with two 500-mL portions of anhydrous ethyl ether. The solid is transferred to a 19 x 10-cm evaporating dish and dried under reduced pressure in a vacuum desiccator to afford crude sodium salt of **1** (270-280 g) as a colorless or slightly tan hygroscopic solid. The material is carried on directly in the continuation of this procedure.

A 4-L Erlenmeyer flask equipped with a 2-in Teflon-coated magnetic stirbar is charged with 1.5 L of deionized water and placed in an ice bath. To the stirring water is added in small portions (*Caution*, Note 7) the crude sodium salt of **1**. When all of the salt has dissolved, the cooled, dark brown solution is acidified to pH 5 (Hydrion paper) by dropwise addition of concd hydrochloric acid. The resultant two-phase mixture is poured into a 3-L separatory funnel and the Erlenmeyer flask is washed with two 100-mL portions of ethyl ether which are added to the separatory funnel. An additional 800 mL of ether is added to the mixture. The upper (organic) and lower (aqueous) phases are separated. The aqueous phase is extracted six additional times (Note 8) with 400 mL of ether each time. The combined organic phases are dried over anhydrous magnesium sulfate, filtered, and concentrated under reduced pressure at 10°C to afford approximately 250 mL of crude **1** as a clear amber oil. This is distilled (Note 9) under reduced pressure (bp 58.5°C, 20 mm) to afford 125-137 g (55-60% yield) of **1** as a slightly yellow oil which solidifies on standing. An analytical sample of **1** was prepared by recrystallization from pentane (mp 42-43°C) (Note 10).

B. A 3-L, three-necked, round-bottomed flask is equipped (see Figure 1) with a 2-in Teflon-coated magnetic stirring bar, a 150-mm Vigreux column topped by a 500-mL pressure-equalizing addition funnel and calcium sulfate drying tube, a short path distillation apparatus, and a ground glass stopper. The flask is charged with 1 L of benzene and 0.5 L of a 60:40 mixture of benzene/methanol (Note 11). To this solution is added 114 g (1.0 mol) of 1-hydroxy-2-methylpent-1-en-3-one (**1**) in a small amount of benzene and 0.95 g (0.005 mol) of p-toluenesulfonic acid monohydrate. The addition funnel is charged with 500 mL of the 60:40 benzene/methanol mixture. The magnetically stirred solution is warmed to a gentle reflux at which time the solution begins to darken to a light brown color. The distillation temperature rises to 59°C (the boiling point of the benzene/methanol azeotrope) and stabilizes. The distillate volume is monitored throughout the course of the reaction. When 250 mL of distillate has collected, the reaction vessel is replenished with 250 mL of fresh benzene/methanol mixture. This cycle is repeated 6 times until starting material is consumed (Note 12). At this time any residual benzene/methanol mixture in the addition funnel is discarded and the funnel is charged with 500 mL of benzene. The reaction volume is maintained by the addition of benzene as the boiling point of the distillate rises to 79°C. In this way, a total of 750 mL of benzene is distilled out of the reaction mixture as the reaction is driven to completion. The reaction mixture is allowed to cool to room temperature and is quenched by the addition of 500 mL of 1.0 M aqueous sodium bicarbonate solution. The two phase mixture is stirred for 5 min and then transferred to a 2-L separatory funnel. The phases are separated and the aqueous (lower) phase is extracted two times with 500 mL of ether each time. The combined organic phases are dried over anhydrous magnesium sulfate, filtered, and concentrated under reduced pressure to give crude **2** as a brown oil. The oil is fractionally distilled under reduced pressure (bp 94-96°C, 22 mm) to afford 90-96 g (70-75%) of **2**. Assay of this material by GLC shows it to be ca. 95% pure (Note 13).

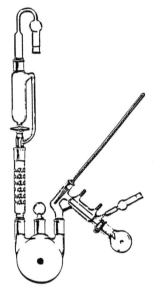

Figure 1

C. A 2-L, round-bottomed flask equipped with a 2-in Teflon-coated magnetic stirring bar is charged with 750 mL of anhydrous ether, 102 g (1.0 mol) of 99% triethylamine (Note 14) and 96 g (0.75 mol) of 1-methoxy-2-methylpent-1-en-3-one (**2**). The flask is capped with a rubber septum equipped with a 16-gauge needle connected to a dry nitrogen source. The magnetically stirred solution is cooled in an ice bath and 167 g (0.75 mol) of trimethylsilyl trifluoromethanesulfonate (Note 15) is added via an 18-gauge cannula over 10 min. The cooled mixture is stirred for 30 min, during which time a red-brown oily precipitate forms. The reaction mixture is transferred to a 2-L separatory funnel and the red-brown (lower) phase is separated and discarded. The remaining ethereal phase is washed with 500 mL of 1.0 M aqueous sodium bicarbonate solution and separated. The aqueous (lower) phase is extracted with 500 mL of ether, separated and discarded. The combined organic phases are dried over anhydrous sodium sulfate, filtered, and concentrated under reduced pressure to yield crude diene **3** as a yellow oil. This is fractionally distilled through a 150-mm Vigreux column at reduced pressure (bp 83-85°C, 22 mm) to afford 121-131 g (81-87%) of **3** as a colorless or slightly yellow liquid. Assay of this material by GLC shows it to be ca. 95% pure (Note 16).

2. Notes

1. Unless otherwise indicated, the solvents are reagent grade and are used without further purification.

2. Commercially available (Aldrich Chemical Company, Inc.) 60% sodium hydride dispersed in mineral oil is dispersed in benzene in the following manner: 81.7 g of 60% sodium hydride mineral oil dispersion is suspended in 100 mL of benzene in a 500-mL Erlenmeyer flask equipped with a Teflon-coated magnetic stirbar. The suspension is vigorously stirred for 1 hr until all of the lumps are gone. The suspension is then transferred to the reaction vessel with a small amount of benzene.

3. Methyl alcohol functions as a catalytic base in this reaction.

4. The efficiency of the procedure is enhanced by complete premixing of the 3-pentanone and ethyl formate.

5. 3-Pentanone and ethyl formate were purchased from Aldrich Chemical Company, Inc. and used without further purification.

6. Efficient mixing is essential during the addition of the ether to ensure the formation of a flocculent precipitate. Inadequate agitation results in the formation of large clumps of product.

7. *Caution: Small amounts of sodium hydride from the previous step may remain in the solid. The addition should be carried out slowly enough to allow any residual sodium hydride to quench completely before the next portion of solid is added.*

8. TLC analysis (Merck 0.25-mm silica gel plates with 254 nm UV indicator, 1:4 ethyl acetate:hexanes) of the organic phase of each extraction reveals the presence of **1** (R_f = 0.32). Extraction is continued until **1** is no longer seen in the organic extracts.

9. Crude **1** may solidify in the pot as any remaining solvent evaporates, but can be reliquified by gentle heating. Stirring is facilitated by a powerful magnetic stirrer and 2-in Teflon-coated magnetic stirbar. To prevent solidification of the distillate on the condenser, the water temperature through the condenser is maintained at 25°C. To minimize product loss, all fractions were collected in a receptacle cooled with an ice bath.

10. The ^{1}H-NMR spectrum of **1** is as follows: (90 MHz, CDCl$_3$, TMS = 0 ppm) δ: 1.13 (t, 3 H, J = 7), 1.79 (s, 3 H), 2.46 (q, 2 H, J = 7), 7.45 (s, 1 H), 7.57 (s, 1 H).

11. The composition of the binary azeotrope of benzene and methanol is 60.5% benzene, 39.5% methanol.

12. The progress of the reaction is monitored by TLC (R_f of **1** = 0.32, R_f of **2** = 0.25; 1:4 ethyl acetate:hexanes). A non-UV absorbing spot (R_f = 0.29) is seen to form as an intermediate between **1** and **2**. This is presumed to be the 1,1-dimethyl acetal of **1**. When this spot is nearly gone, the reaction is presumed to be complete.

13. The ^{1}H-NMR of **2** is as follows: (90 MHz, CDCl$_3$, TMS) δ: 1.10 (t, 3 H, J - 7.4), 1.71 (s, 3 H), 2.52 (q, 2 H, J = 7.4), 3.86 (s, 3 H), 7.23 (bs, 1 H).

14. Triethylamine was purchased from Aldrich Chemical Company, Inc. and was used without further purification.

15. Trimethylsilyl trifluoromethanesulfonate was purchased from Petrarch Systems and was used without further purification.

16. The ^{1}H-NMR of **3** is as follows: (90 MHz, CDCl$_3$, CHCl$_3$, δ 7.27) δ: 0.23 (s, 9 H), 1.63 (d, 3 H, J = 7), 1.70 (s, 3 H), 3.66 (s, 3 H), 4.75 (q, 1 H, J = 7), 6.35 (s, 1 H).

Waste Disposal Information

All toxic materials were disposed of in accordance with "Prudent Practices for Disposal of Chemicals from Laboratories"; National Academy Press; Washington, DC, 1983.

3. Discussion

Siloxy dienes have been shown to be highly effective in both the Diels-Alder[3] reaction and the hetero Diels-Alder (diene-aldehyde cyclocondensation)[4] reactions. Diene **3** has been used in the synthesis of several important natural products including zincophorin,[5] rifamycin,[6] and the Prelog-Djerassi lactone.[7] Recently, under the aegis of chiral catalysts, **3** has been shown to participate in the diene-aldehyde cyclocondensation reaction with several aldehydes to afford cycloadducts of very high enantiomeric excess.[8]

The procedure described here allows the convenient preparation of large quantities of diene **3** and has several advantages over previously published sequences. 1-Methoxy-2-methylpenten-3-one (**2**) can be prepared in ca. 90-g lots and can be stored for several months at 0°C under argon. The earlier procedure for the synthesis of **2**[9] relies on dimethyl sulfate as the methylating reagent in the formation of the methyl enol ether moiety. The high toxicity of this reagent renders this strategy unattractive for routine use. The published procedure for the conversion of **2** to 1-methoxy-2-methyl-3-(trimethylsiloxy)-1,3-pentadiene (**3**)[3a] uses trimethylsilyl chloride, triethylamine and a catalytic amount of zinc chloride in benzene (40°C, ca. 12

hr) for the silylation. This mixture results in the formation of copious quantities of triethylamine hydrochloride, which greatly complicates the work up and purification. This new procedure, using trimethylsilyl trifluoromethanesulfonate and triethylamine, allows the convenient separation of the amine salt and significantly shortens the reaction time (30 min vs 12 hr). The silylation can be carried out on scales ranging from 0.10 mol to 0.75 mol with consistently good yields. Diene **3** can be stored for several months at 0°C under argon without significant decomposition.

1. Department of Chemistry, Yale University, New Haven, CT 06511.
2. The submitters would like to thank Professor Samuel J. Danishefsky and Dr. Sarah E. Danishefsky for their support and guidance.
3. (a) Danishefsky, S.; Yan, C.-F.; Singh, R. K.; Gammill, R. B.; McCurry, Jr., P.M.; Fritsch, N.; Clardy, J. *J. Am. Chem. Soc.* **1979**, *101*, 7001, and references therein; (b) Danishefsky, S. *Acc. Chem. Res.* **1981**, *14*, 400.
4. Danishefsky, S. J. *Aldrichimica Acta* **1986**, *19*, 59.
5. Danishefsky, S. J.; Selnick, H. G.; DeNinno, M. P.; Zelle, R. E. *J. Am. Chem. Soc.* **1987**, *109*, 1572.
6. Danishefsky, S.J.; Myles, D. C.; Harvey, D. F. *J. Am. Chem. Soc.* **1987**, *109*, 862.
7. Danishefsky, S. J.; Larson, E.; Askin, D.; Kato, N. *J. Am. Chem. Soc.* **1985**, *107*, 1246, and references therein.
8. (a) Maruoka, K.; Itoh, T.; Shirasaka, T.; Yamamoto, H. *J. Am. Chem. Soc.* **1988**, *110*, 310; (b) Bednarski, M.; Danishefsky, S. *J. Am. Chem. Soc.* **1986**, *108*, 7060.
9. Sugasawa, S.; Yamada, S.-i.; Narahashi, M. *J. Pharm. Soc. Japan* **1951**, *71*, 1345; *Chem. Abstr.* **1952**, *46*, 8034d.

Appendix

Chemical Abstracts Nomenclature (Collective Index Number); (Registry Number)

(E,Z)-1-Methoxy-2-methyl-3-(trimethylsiloxy)-1,3-pentadiene: Silane, [[1-(2-methoxy-1-methylethenyl)-1-propenyl]oxy]trimethyl-, (Z,E)- (10); (72486-93-2)

Sodium hydride (8,9); (7646-69-7)

3-Pentanone (8,9); (96-22-0)

Ethyl formate: Formic acid, ethyl ester (8,9); (109-94-4)

1-Hydroxy-2-methylpent-1-en-3-one: 1-Penten-3-one, 1-hydroxy-2-methyl- (9); (50421-81-3)

p-Toluenesulfonic acid monohydrate (8); Benzenesulfonic acid, 4-methyl-, monohydrate (9); (6192-52-5)

Trimethylsilyl trifluoromethanesulfonate: Methanesulfonic acid, trifluoro-, trimethylsilyl ester (8,9); (27607-77-8)

NICKEL-CATALYZED SILYLOLEFINATION OF ALLYLIC DITHIOACETALS: (E,E)-TRIMETHYL(4-PHENYL-1,3-BUTADIENYL)SILANE

(Silane, trimethyl (4-phenyl-1,3-butadienyl)-, (E,E)-)

A. Ph-CH=CH-CHO + HSCH$_2$CH$_2$SH $\xrightarrow{BF_3 \cdot OEt_2}$ Ph-CH=CH-CH(SCH$_2$CH$_2$S)

B. Me$_3$SiCH$_2$Cl + Mg $\longrightarrow$ Me$_3$SiCH$_2$MgCl

C. Ph-CH=CH-CH(SCH$_2$CH$_2$S) + Me$_3$SiCH$_2$MgCl $\xrightarrow{NiCl_2(PPh_3)_2}$ Ph-CH=CH-CH=CH-SiMe$_3$

Submitted by Zhi-Jie Ni and Tien-Yau Luh.[1]
Checked by Shigeki Habaue and Hisashi Yamamoto.

1. Procedure

Caution! 1,2-Ethanedithiol has a powerful stench. Steps A and C should be performed in a well-ventilated hood.

A. *2-(2-Phenylethenyl)-1,3-dithiolane.* In a 1-L, round-bottomed flask equipped with a magnetic stirring bar are placed 26.8 g (0.2 mol) of freshly distilled (E)-3-phenyl-2-propenal and 20.1 g (0.21 mol) of 1,2-ethanedithiol in 400 mL of chloroform. To the stirred solution is added 10 mL (11.3 g, 0.080 mol) of boron trifluoride etherate in one portion. The mixture is stirred at room temperature for 2 hr. The chloroform solution is washed with two 100-mL portions of 10% aqueous sodium hydroxide. The aqueous layer is extracted twice with 100 mL of chloroform. The combined organic layers are washed twice with 200 mL of water, dried over anhydrous magnesium sulfate, and filtered. The filtrate is concentrated under reduced pressure to give 40.6 g (97%) of a white solid which is sufficiently pure for the next operation.

B. *(Trimethylsilyl)methylmagnesium chloride.* A 500-mL, three-necked, round-bottomed flask containing 5.2 g (0.22 g-atom) of magnesium turnings is equipped with a rubber septum, a reflux condenser, an addition funnel and a magnetic stirring bar. The system is flame-dried and flushed with nitrogen. A few crystals of iodine and 150 mL of anhydrous ether (Note 1) are introduced. As the contents of the flask are stirred, 25.8 g (0.21 mol) of (chloromethyl)trimethylsilane (Note 2) is added in small portions until the reaction begins, and then at such a rate as to maintain gentle refluxing of the ether. The addition requires about 30 min, after which the mixture is heated under reflux for an additional 30 min. The solution is cooled to room temperature and is used directly for the next reaction.

C. *(E,E)-Trimethyl(4-phenyl-1,3-butadienyl)silane.* In a 1-L, two-necked, round-bottomed flask fitted with a reflux condenser, rubber septum, and a magnetic stirring bar are placed 14.6 g (0.070 mol) of 2-(2-phenylethenyl)-1,3-dithiolane and 2.3 g (0.0035 mol) of dichlorobis(triphenylphosphine)nickel (Note 3). The flask is evacuated and flushed with nitrogen three times. To the above mixture is added 200 mL of anhydrous tetrahydrofuran (Note 4); then it is cooled in an ice bath. The ether solution of (trimethylsilyl)methylmagnesium chloride prepared above is introduced with a double-ended needle in one portion (Note 5). The mixture is refluxed for 10 hr, cooled to room temperature, and treated with 200 mL of saturated ammonium chloride solution. The organic layer is separated and the aqueous layer is extracted with three, 200-mL portions of ether. The combined organic layers are washed twice with 100 mL of aqueous 10% sodium hydroxide solution and twice with 100 mL of brine. The organic solution is dried over anhydrous magnesium sulfate. The solvent is removed under reduced pressure and the residue is filtered through a short column packed with 30 g of silica gel (Note 6) and flushed under a positive nitrogen pressure with 300 mL of hexane. After evaporation of the solvent under reduced pressure, the yellowish residue is distilled to give 12.9 g (91%) of (E,E)-trimethylsilyl(4-phenyl-1,3-butadienyl)silane (Note 7) as a colorless liquid, bp 99-101°C (0.6 mm), which solidifies on standing, mp <37°C.

2. Notes

1. Ethyl ether is distilled from sodium-benzophenone ketyl before use.
2. (Chloromethyl)trimethylsilane, also available from Aldrich Chemical Company, Inc., was purchased from Wako Pure Chemical Industries, LTD. and used directly.
3. Dichlorobis(triphenylphosphine)nickel, also available from Aldrich Chemical Company, Inc., was purchased from Fluka AG, and used without further purification. The catalyst can also be prepared according to literature procedures.[2]
4. Tetrahydrofuran is distilled from sodium-benzophenone ketyl before use.
5. An excess of the Grignard reagent is required to maximize the yield; otherwise, the reaction is incomplete.
6. Silica gel (230-400 mesh) was purchased from E. Merck Co.
7. The spectral properties of the product are as follows: IR (neat) cm^{-1}: 3040, 1605, 1450, 1250, 1000, 870, 840, 730, 690; ^{1}H NMR (CDCl$_3$) δ: 0.13 (s, 9 H, -Si(CH$_3$)$_3$), 6.01 (d, 1 H, J = 17.6, =CHTMS), 6.58 (d, 1 H, J = 15.2, PhCH=), 6.69 (dd, 1 H, J = 9.8 and 17.6, -CH=CHTMS), 6.83 (dd, 1 H, J = 9.8 and 15.2, PhCH=CH-), 7.20-7.40 (m, 5 H, aromatic) (The assignments for olefinic protons are based on simulation results); ^{13}C NMR (CDCl$_3$) δ: -1.5, 126.7, 127.8, 128.7, 131.8, 133.0, 135.2, 137.4, 144.3; exact mass calcd for C$_{13}$H$_{18}$Si: 202.1178; found 202.1185.

Waste Disposal Information

All toxic materials were disposed of in accordance with "Prudent Practices for Disposal of Chemicals from Laboratories"; National Academy Press; Washington, DC, 1983.

3. Discussion

Trimethyl(4-phenyl-1,3-butadienyl)silane can be prepared by the reaction of bis(trimethylsilyl)methyllithium with cinnamaldehyde[3] or 1,3-bis(trimethylsilyl)propenyllithium with benzaldehyde.[4,5] The reaction of 1-bromo-2-phenylthioethene with (E)-2-trimethylsilylethenylmagnesium bromide in the presence of a palladium catalyst gives the corresponding dienyl sulfide which serves as a precursor for the preparation of butadienylsilanes.[6] However, the starting materials in most of these syntheses are not readily available.

The procedure described here is based on a series of reports on the nickel-catalyzed coupling reactions of dithioacetals with Grignard reagents.[7-11] The method offers a new, very efficient and convenient route to the substituted butadienylsilanes.[7] The starting materials are easily accessible and the operation is very simple. The reaction in general is highly stereoselective, if not stereospecific. The phenyl substituent can be replaced with simple alkyl groups and yields essentially remain unchanged.[7] The extension of this method to the synthesis of trienylsilanes has proved successful.[7] Since vinylsilanes can be converted into α,β-unsaturated aldehydes,[12-16] the combination of these latter procedures with the method described here can be used in the homologation of enals. (E)-β-Arylvinylsilanes are synthesized stereospecifically in a similar manner from benzylic dithioacetals.[8]

1. Department of Chemistry, The Chinese University of Hong Kong, Shatin, N. T., Hong Kong, and Department of Chemistry, National Taiwan University, Taipei, Taiwan, Republic of China.
2. (a) Cotton, F. A.; Faut, O. D.; Goodgame, D. M. L. *J. Am. Chem. Soc.* **1961**, *83*, 344; (b) Venanzi, L. M. *J. Chem. Soc.* **1958**, 719.
3. Carter, M. J.; Fleming, I.; Percival, A. *J. Chem. Soc., Perkin Trans. I* **1981**, 2415.
4. Chan, T.-H.; Li, J.-S. *J. Chem. Soc., Chem. Commun.* **1982**, 969.
5. Corriu, R.; Escudie, N.; Guerin, C. *J. Organometal. Chem.* **1984**, *264*, 207.
6. Fiandanese, V.; Marchese, G.; Mascolo, G.; Naso, F.; Ronzini, L. *Tetrahedron Lett.* **1988**, *29*, 3705.
7. Ni, Z.-J.; Luh, T.-Y. *J. Org. Chem.* **1988**, *53*, 5582.
8. Ni, Z.-J.; Luh, T.-Y. *J. Org. Chem.* **1988**, *53*, 2129.
9. Ni, Z.-J.; Luh, T.-Y. *J. Chem. Soc., Chem. Commun.* **1987**, 1515,
10. Ni, Z.-J.; Luh, T.-Y. *J. Chem. Soc., Chem. Commun.* **1988**, 1011.
11. Yang, P.-F.; Ni, Z.-J.; Luh, T.-Y. *J. Org. Chem.* **1989**, *54*, 2261.
12. Chan, T. H.; Lau, P. W. K.; Mychajlowskij, W. *Tetrahedron Lett.* **1977**, 3317.
13. Pillot, J. P.; Dunogues, J.; Calas, R. *Bull. Soc. Chim. Fr.* **1975**, 2143.
14. Yamamoto, K.; Nunokawa, O.; Tsuji, J. *Synthesis* **1977**, 721.
15. Yamamoto, K.; Yoshitake, J.; Nguyen Thi Qui; Tsuji, J. *Chem. Lett.* **1978**, 859.
16. Yamamoto, K.; Ohta, M.; Tsuji, J. *Chem. Lett.* **1979**, 713.

Appendix

Chemical Abstracts Nomenclature (Collective Index Number); (Registry Number)

(E,E)-Trimethyl(4-phenyl-1,3-butadienyl)silane: Silane, trimethyl(4-phenyl-1,3-butadienyl)-, (E,E)- (10); (70960-88-2)

1,2-Ethanediol (8,9); (540-63-6)

2-(2-Phenylethenyl)-1,3-dithiolane: 1,3-Dithiolane, 2-styryl- (8); 1,3-Dithiolane, 2-(2-phenylethenyl)- (9); (5616-58-0)

(E)-3-Phenyl-2-propenal: Cinnamaldehyde (E)- (8); 2-Propenal, 3-phenyl-, (E)-(9); (14317-10-9)

Boron trifluoride etherate: Ethyl ether, compd. with boron fluoride (1:1) (8); Ethane, 1,1'-oxybis-, compd. with trifluoroborane (1:1) (9); (109-63-7)

(Trimethylsilyl)methylmagnesium chloride: Magnesium, chloro[(trimethylsilyl)methyl]- (8,9); (13170-43-9)

Chloromethyltrimethylsilane: Silane, (chloromethyl)trimethyl- (8,9); (2344-80-1)

Dichlorobis(triphenylphosphine)nickel: Nickel, dichlorobis(triphenylphosphine)- (8,9); (14264-16-5)

α-ACETYLENIC ESTERS FROM α-ACYLMETHYLENEPHOSPHORANES: ETHYL 4,4,4-TRIFLUOROTETROLATE

(2-Butynoic acid, 4,4,4-trifluoro-, ethyl ester)

A. [Ph$_3$P$^+$—CH$_2$—CO$_2$Et] Br$^-$ $\xrightarrow[\text{(CF}_3\text{CO)}_2\text{O}]{\text{2 eq. Et}_3\text{N}}$ Ph$_3$P=C(CO$_2$Et)(COCF$_3$)

B. Ph$_3$P=C(CO$_2$Et)(COCF$_3$) $\xrightarrow[\text{1 - 2 torr}]{\text{150° - 200°C}}$ CF$_3$—C≡C—CO$_2$Et

Submitted by B. C. Hamper.[1]
Checked by T. Harrison and Larry E. Overman.

1. Procedure

A. *Ethyl 4,4,4-trifluoro-2-(triphenylphosphoranylidene)acetoacetate.* A 2-L, four-necked, round-bottomed flask is equipped with a nitrogen line attached to a bubbler, a 250-mL pressure-equalizing funnel, an overhead stirrer and a thermometer. The flask is charged with 215 g (0.5 mol) of (carbethoxymethyl)triphenylphosphonium bromide and 1.1 L of anhydrous tetrahydrofuran (THF) (Notes 1 and 2). The stirred suspension is cooled in an ice water bath and treated with 150 mL (1.1 mol) of triethylamine (Note 3) added dropwise over 5 min. After the mixture is stirred for an additional 30 min at 5°C, it is treated dropwise with 78 mL (116 g, 0.55 mol) of trifluoroacetic anhydride (Note 4) in such a manner that the reaction temperature is maintained between 5-10°C, which results in a total addition time of approximately 1 hr. The mixture is allowed to stir for 2 hr and subsequently filtered, the precipitate is washed three times with cold THF and the filtrate is concentrated under reduced pressure to afford a yellow oily residue. Trituration of the residue with 600 mL of water affords a crystalline product which is collected, washed three times with 100 mL of water, and dried by suction to afford 208 g of a yellowish colored solid (Note 5). The solid is dissolved in 900 mL of hot methanol, filtered, the solution treated with 500 mL of water and placed in a refrigerator overnight. The crystalline product is collected, washed three times with 100 mL of cold water and dried under reduced pressure to afford 200-208 g (89-93%) of ethyl 4,4,4-trifluoro-2-(triphenylphosphoranylidene)-acetoacetate as a very pale-yellow, crystalline solid (Note 6).

B. *Ethyl 4,4,4-trifluorotetrolate.* A 1-L, three-necked, round-bottomed flask is equipped with an efficient magnetic stirrer, a heating mantel, a thermocouple (Note 7) and a large bore gas exit tube (Figure 1, Note 8). To this flask are added 200 g (0.45 mol) of ethyl 4,4,4-trifluoro-2-(triphenylphosphoranylidene)acetoacetate and 40 g of potassium carbonate (Note 9). The gas exit tube is connected to a Dewar condenser, modified with a side arm connection, which has a round-bottomed flask for collection of the product in a dry ice-acetone bath. A second trap is placed between the reaction setup and the vacuum pump, all the traps are cooled with dry ice-acetone, and the system is evacuated to 1-2 mm. The reaction flask is carefully heated to 150°C (Note 10) at which point the phosphorane melts and the acetylene evolution begins. The molten phosphorane is stirred and heated from 160°C to 220°C over a period of 5 hr. Heating is carefully increased during the reaction in order to control the rate of distillation of the acetylenic product. From the round-bottomed flask in the cold trap is obtained 65-67 g (87-89%) of a clear, slightly yellow, liquid. The thermolysis product is dried with magnesium sulfate and filtered through Celite. Distillation at atmospheric pressure through a short-path distillation apparatus affords about 1.0 g of forerun and

59-61 g (79-82%) of ethyl 4,4,4-trifluorotetrolate as an analytically pure, colorless liquid, bp$_{760}$ 97-100°C (Note 11).

2. Notes

1. Anhydrous tetrahydrofuran was obtained from Aldrich Chemical Company, Inc., in SureSeal bottles and was used without further purification. (Carbethoxymethyl)triphenylphosphonium bromide may be obtained from Aldrich Chemical Company Inc., or prepared as described in *Org. Synth., Coll. Vol. VII* **1990**, 232. The in situ preparation described in Note 2 was used by the checkers.

2. We have found it more convenient to prepare (carbethoxymethyl)-triphenylphosphonium bromide in situ in the same reaction vessel from triphenylphosphine and ethyl bromoacetate. This effectively provides a one-pot preparation for the α-acylmethylenephosphorane from triphenylphosphine and allows facile incorporation of various esters, by employing different bromoacetate esters, in the acetylenic product. The yield and purity of the resultant phosphorane is unaffected by the following one-pot procedure.

A 2-L, four-necked, round-bottomed flask is equipped with a nitrogen line attached to a bubbler, a 250-mL pressure-equalizing funnel, an overhead stirrer and a thermometer. The flask is charged with 132 g of triphenylphosphine (0.503 mol) and 500 mL of anhydrous tetrahydrofuran and cooled in an ice water bath to 5°C. The stirred solution is treated dropwise with 56 mL of ethyl bromoacetate (84 g, 0.503 mol) added at such a rate that the temperature is maintained between 8°C and 10°C. The total addition time is about 15 min. After the mixture is stirred overnight, it is diluted with an additional 600 mL of anhydrous tetrahydrofuran and the precipitate is washed from the sides of the flask. The resultant suspension of (carbethoxymethyl)triphenylphosphonium bromide is cooled in an ice-water bath and used directly in the same pot to prepare α–acylmethylenephosphoranes as detailed in section A.

3. Triethylamine was supplied by Eastman Kodak Company and Fisher Scientific Company.

4. Trifluoroacetic anhydride was obtained from Aldrich Chemical Company, Inc.

5. The yellow crystalline solid is analytically pure (>95% by NMR and HPLC analysis), mp 124-127°C. We have found that the yield of the acetylenes can be adversely affected by small amounts of impurities in the α–acylmethylenephosphoranes and prefer to recrystallize the phosphoranes prior to thermolysis. The yield from the recrystallization step is greater than 95%.

6. The phosphorane softens above 120°C and melts between 125-130°C (lit.[2,3] mp 125-127°C). Spectral properties of the phosphorane are as follows: ^{1}H NMR (500 MHz, CDCl$_3$) δ: 0.87 (t, 3 H, J = 7.2), 3.81 (q, 2 H, J = 7.2), 7.49 (dt, 6 H, J = 7.9, 3.3), 7.57 (m, 3 H), 7.67 (dd, 6 H, J = 12.9, 7.9); ^{13}C NMR (125 MHz, CDCl$_3$) δ: 13.5, 59.8, 70.1 (d, $^1J_{CP}$ = 110), 117.9 (dd, $^1J_{CF}$ = 288, $^3J_{CP}$ = 14.6), 123.7 (d, $^1J_{CP}$ = 93.3), 128.8 (d, $^2J_{CP}$ = 13.0), 132.2 (d, $^4J_{CP}$ = 3.0), 133.2 (d, $^3J_{CP}$ = 10.0), 165.6 (d, $^2J_{CP}$ = 13.0), 174.4 (dd, $^2J_{CF}$ = 34.0, $^2J_{CP}$ = 5.8); ^{31}P NMR (202 MHz, CDCl$_3$) δ: 19.8. Anal. Calcd for $C_{24}H_{20}O_3F_3P$: C, 64.87; H, 4.54. Found: C, 64.96; H, 4.60.

7. A stainless steel-covered thermocouple is preferred since it is less likely to break than a mercury thermometer. As the phosphorane begins to melt, large chunks of solid remain as the mixture is initially stirred, and these can lodge between a thermometer and the walls of the flask. The thermocouple has the added advantage that the steel rod can be used to help break up the melting phosphorane, taking care to avoid breaking the flask.

8. The large bore gas exit tube (Figure 1) used between the reaction flask and the Dewar condenser is a glass tee equipped with the appropriate 24/40 ground glass joints. Alternatively, the connection can be made using large bore, thick wall Tygon or vacuum tubing.

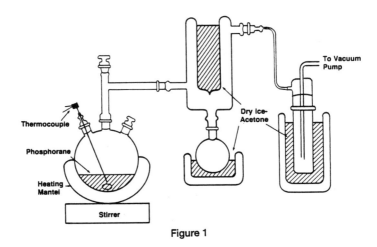

Figure 1

9. In the absence of potassium carbonate, the thermolysis product is slightly acidic and the pH of a water wash of freshly collected product is about 1.

10. *Caution! Care must be taken not to heat the phosphorane too rapidly, particularly before the solid has melted. The temperature of the molten material is not at equilibrium until the mixture has completely melted. If it is heated too rapidly, it is difficult to control the rate of acetylene production. It is best to heat slowly until a stirred, molten material is obtained and to continue heating at such a rate as to control acetylene formation.* We have obtained excellent results applying initially 30 volts to a 380-watt, 115-volt, 1-L heating mantel obtained from Glas-Col Apparatus Company. Alternatively, more even and controlled heating can be obtained by using a large oil bath.

11. Previous literature reports[3] bp 96-98°C. Spectral properties of the acetylene are as follows: ^{1}H NMR (500 MHz, CDCl$_3$) δ: 1.34 (t, 3 H, J = 7.1), 4.32 (q, 2H, J = 7.1); ^{13}C NMR (125 MHz, CDCl$_3$) δ: 13.8, 63.5, 69.9 (q, $^2J_{CF}$ = 54.0), 75.5 (q, $^3J_{CF}$ = 6.0), 113.4 (q, $^1J_{CF}$ = 259), 150.7; IR (neat) cm^{-1}: 2987, 2275 (w); 1731. Anal. Calcd for C$_6$H$_5$O$_2$F$_3$: C, 43.49; H, 3.03. Found: C, 43.14; H, 3.07.

Waste Disposal Information

All toxic materials were disposed of in accordance with "Prudent Practices for Disposal of Chemicals from Laboratories"; National Academy Press; Washington, DC, 1983.

3. Discussion

Thermolysis of α–acylmethylenephosphoranes is the most convenient method for preparation of acetylenes with perfluoroalkyl substituents.[3,4] Many of these acetylenes, particularly the trifluoromethyl analogs, are particularly volatile and thermolysis allows preparation and collection of material which is free of solvents or impurities of similar boiling point. In addition, the acetylenes are prepared from readily available starting materials with overall conversions for the two steps of greater than 75% in many cases. Attempts to prepare trifluorotetrolic acid by carbonylation of the lithium acetylenide of 3,3,3-trifluoropropyne have been unsuccessful.[5] However, the treatment of lithium acetylenides with chloroformates affords perfluoroalkyl-substituted propiolates (the perfluoroalkyl group is a C$_4$ chain length or longer) in 38-43% yield along with an unusual by-product.[6] Benzyl trifluorotetrolate, which was employed in the synthesis of a trifluoromethyl analog of geraniol, has been prepared from ethyl trifluoroacetoacetate via a pyrazolone in an overall yield of 55%.[7] A number of (difluoroalkyl)propiolates, employed in Diels-Alder cycloaddition reactions, have been prepared by treatment of the corresponding ketones with diethylaminosulfur trifluoride (DAST).[8] The corresponding (difluoroalkyl)propiolic acids have also been prepared by carbonylation of the magnesium bromoacetylenide.[9]

The utility of the phosphorane route to electron-deficient acetylenes depends on the facile synthesis of the α-acylmethylenephosphorane intermediates. Previously they had been prepared by a two-step procedure from the available phosphonium salts, requiring isolation of the Wittig reagent [e.g., (ethoxycarbonylmethylene)-triphenylphosphorane] intermediate.[3,4] Acylation of the Wittig reagent affords, via transylidation,[10] a 1:1 mixture of components which must be separated prior to thermolysis. In addition, at least half of the Wittig reagent is lost by conversion to the starting phosphonium salt. The addition of a suitable base, such as triethylamine, to the reaction mixture avoids the undesired transylidation reaction and affords complete conversion to the desired α-acylmethylenephosphorane. For acyl halides which have an acidic hydrogen α to the carbonyl group, treatment with base can give rise to ketenes which readily react with Wittig reagents to afford allenes.[11] This route, using triethylamine as a base, has been explored to prepare allenes that are either substituted or unsubstituted in the α-position.

For acyl halides or anhydrides which do not afford ketenes in the presence of base (such as perfluoroacyl halides), however, the α-acylmethylenephosphoranes can be prepared directly in one step from the phosphonium salts by using two equivalents of base by the present procedure (Table I).[2] Both tetrahydrofuran and methylene chloride have been used as solvents and in the case of the title compound, tetrahydrofuran provides the best results. Good yields of the phosphoranes are generally obtained when R^1 is an electron-withdrawing group such as ester or nitrile. The yields of phosphoranes obtained for the thiomethyl or phenyl cases can be improved by using 1,4-diazabicyclo[2.2.2]octane (DABCO) rather than triethylamine as the base.

1. Monsanto Agricultural Company, A Unit of Monsanto Company, 800 N. Lindbergh Blvd., St. Louis, MO 63167.
2. Hamper, B. C. *J. Org. Chem.* **1988**, *53*, 5558.
3. Huang, Y.; Shen, Y.; Xin, Y.; Wang, Q.; Wu, W. *Sci. Sin. (Engl. Ed.)* **1982**, *25*, 21.
4. For some recent examples, see Shen, Y.; Zheng, J.; Huang, Y. *J. Fluorine Chem.* **1988**, *41*, 363; Braga, A. L.; Comasseto, J. V.; Petragnani, N. *Tetrahedron Lett.* **1984**, *25*, 1111; Kobayashi, Y.; Yamashita, T.; Takahashi, K.; Kuroda, H.; Kumadaki, I. *Chem. Pharm. Bull.* **1984**, *32*, 4402; Shen, Y.; Lin, Y.; Xin, Y. *Tetrahedron Lett.* **1985**, *26*, 5137.
5. Braga, A. L.; Comasseto, J. V.; Petragnani, N. *Synthesis* **1984**, 240.
6. Froissard, J.; Greiner, J.; Pastor, R.; Cambon, A. *J. Fluorine Chem.* **1981**, *17*, 249; Chauvin, A.; Greiner, J.; Pastor, R.; Cambon, A. *J. Fluorine Chem.* **1984**, *25*, 259.
7. Poulter, C. D.; Wiggins, P. L.; Plummer, T. L. *J. Org. Chem.* **1981**, *46*, 1532.
8. Hirao, K.; Yamashita, A.; Yonemitsu, O. *J. Fluorine Chem.* **1987**, *36*, 293.
9. Yamanaka, H.; Araki, T.; Kuwabara, M.; Fukunishi, K.; Nomura, M. *Nippon Kagaku Kaishi*, **1986**, 1321; *Chem. Abstr.* **1987**, *107*, 175447f.
10. Bestmann, H. J. *Chem. Ber.* **1962**, *95*, 58.
11. Lang, R. W.; Hansen, H.-J. *Helv. Chim. Acta* **1980**. *63*, 438; Lang, R. W.; Hansen, H.-J. *Org. Synth., Coll. Vol. VII* **1990**, 232.

TABLE I
PREPARATION OF (α–ACYLMETHYLENE)PHOSPHORANES FROM α–SUBSTITUTED METHYLPHOSPHONIUM SALTS[2]

$$Ph_3P^+\text{-}CH(R^1)\ Br^- \xrightarrow[(R^2CO)_2O]{\text{2 eq. base}} Ph_3P=C(R^1)\text{-}C(=O)\text{-}R^2$$

R^1	R^2	Solvent	Base	Yield (%)
CO$_2$Et	CF$_3$	CH$_2$Cl$_2$	Et$_3$N	54
CO$_2$Me	CF$_3$	THF	Et$_3$N	99
CO$_2$Et	CF$_2$CF$_3$	THF	Et$_3$N	69
CN	CF$_3$	THF	Et$_3$N	75
SCH$_3$	CF$_3$	THF	Et$_3$N	49
Ph	CF$_3$	CH$_2$Cl$_2$	Et$_3$N	24
Ph	CF$_3$	CH$_2$Cl$_2$	DABCO	57

Appendix

Chemical Abstracts Nomenclature (Collective Index Number); (Registry Number)

Ethyl 4,4,4-trifluorotetrolate: 2-Butynoic acid, 4,4,4-trifluoro-, ethyl ester (10); (79424-03-6)

Ethyl 4,4,4-trifluoro-2-(triphenylphosphoranylidene)acetoacetate: Butanoic acid, 4,4,4-trifluoro-3-oxo-2-(triphenylphosphoranylidene)-, ethyl ester (11); (83961-56-2)

(Carbethoxymethyl)triphenylphosphonium bromide: Phosphonium, (carboxymethyl)triphenyl-, bromide, ethyl ester (8); Phosphonium, (2-ethoxy-2-oxoethyl)triphenyl-, bromide (9); (1530-45-6)

Trifluoroacetic anhydride: Acetic acid, trifluoro-, anhydride (8,9); (407-25-0)

Triphenylphosphine: Phosphine, triphenyl- (8,9); (603-35-0)

Ethyl bromoacetate: Acetic acid, bromo-, ethyl ester (8,9); (105-36-2)

SYNTHESIS OF β-KETO ESTERS BY C-ACYLATION OF PREFORMED ENOLATES WITH METHYL CYANOFORMATE: PREPARATION OF METHYL (1α,4aβ,8aα)-2-OXODECAHYDRO-1-NAPHTHOATE

Submitted by Simon R. Crabtree, Lewis N. Mander, and S. Paul Sethi.[1]
Checked by Thierry Vandenheste and Leo A. Paquette.

1. Procedure

Caution! Cyanide salts are formed in this procedure. All procedures should be conducted in a well-ventilated hood and rubber gloves should be worn.

Liquid ammonia (350 mL) is dried over sodium amide for 20 min (Note 1), then distilled under a positive pressure of dry nitrogen into a flame-dried, 1-L, three-necked, round-bottomed flask equipped with a dry ice condenser, rubber septum and magnetic stirring bar. The flask is cooled to -78°C and small pieces of lithium (1.11 g, 0.16 mol) (Note 2) are added over 10 min to afford a deep blue solution. A solution of enone 1 (10.08 g, 0.07 mol) (Note 3) in tert-butyl alcohol (5.10 g, 0.07 mol) (Note 4) and ether (40 mL) (Note 5) is added dropwise over a 15-min period at -78°C, and then sufficient isoprene (Note 6) is added dropwise to discharge the residual blue color of the reaction mixture. The ammonia is allowed to evaporate under a stream of dry nitrogen and the ether is removed under reduced pressure to leave a white foam. After a further 5 min under high vacuum, the nitrogen atmosphere is restored, the lithium enolate **2** is suspended in dry ether (80 mL) at -78°C by vigorous stirring, and methyl cyanoformate (6.73 g, 0.08 mol) (Note 7) is added over a 5-min period. Stirring is maintained at this temperature for a further 40 min and then the mixture is allowed to warm to 0°C. Water (500 mL) and ether (200 mL) are added and the mixture is stirred vigorously until the precipitate has dissolved. After separation of the organic layer, the aqueous phase is extracted with ether (2 x 300 mL), the combined extracts are washed with water (200 mL) and brine (200 mL) (Note 8), and then dried over anhydrous magnesium sulfate. Removal of solvent under reduced pressure furnishes a pale yellow oil which is dissolved in hexane (200 mL); the solution is kept in the freezer (-28°C) for approximately 1.5 hr, after which time very pale yellow crystals of the β-keto ester **3** are deposited. These are collected by filtration at low temperature and after concentration of the mother liquors, two further crops are collected. Flash chromatography (ethyl acetate/hexane, 13:87) on silica gel (140 g) (Note 9) of the residue (3.5 g) from the remaining mother liquor followed by two recrystallizations affords further **3**. The total quantity of **3** obtained is 11.5-11.8 g (81-84% yield) (Note 10).

2. Notes

1. Sodium amide is generated in situ from sodium metal and a crystal of ferric nitrate. Sodium metal by itself is not an effective drying agent.

2. Lithium wire was purchased from Merck Schuchardt.

3. Enone **1** was prepared according to the procedure of Augustine, R. L.; Caputo, J. A. *Org. Synth., Coll. Vol. V* **1973**, 869.

4. tert-Butyl alcohol was purchased from Ajax Chemicals and was freshly distilled from calcium hydride.

5. Diethyl ether was purchased from Ajax Chemicals and was freshly distilled from sodium and benzophenone.

6. Isoprene was purchased from Eastman Kodak, Ltd.

7. Methyl cyanoformate was purchased from the Aldrich Chemical Company, Inc. The ethyl and benzyl analogues are also available. Small quantities up to 30 g are conveniently prepared by the method of Childs and Weber,[2] but several workers have found it to be unsatisfactory on a larger scale because of the generation of inseparable impurities.

8. The aqueous washings contain toxic lithium cyanide and should be treated with a strong oxidizing agent (e.g., potassium permanganate) before disposal.

9. Merck Kieselgel 60 silica gel was used.

10. A sample recrystallized twice from hexane had mp 48-49°C (Anal. Calcd for $C_{12}H_{18}O_3$: C, 68.53; H. 8.63%; Found C, 68.59; H, 8.93). The spectroscopic properties of this product are as follows: IR (CHCl$_3$) cm^{-1}: 3026, 2931, 2859, 1744, 1710, 1454, 1447, 1437, 1364, 1348, 1340, 1155, 1112; ^{1}H NMR (CDCl$_3$) δ: 1.00-1.50 (m, 6 H), 1.65-1.8 (m, 5 H), 1.9-2.0 (m, 1 H CH), 2.30-2.50 (m, 2 H, CH$_2$), 3.12 (d, 1 H, J = 12), 3.76 (s, 3 H, OMe). ^{13}C NMR (CDCl$_3$) δ: 25.4, 25.9, 32.6, 32.9, 33.3, 40.5, 41.4, 45.5, 51.8, 53.8, 170.2, 205.6; Mass spectrum (m/z): 210 (M+, 34.6%), 192 (72), 122 (66), 108 (59), 95 (68), 94 (94), 81 (100), 79 (53), 67 (77), 51 (61).

Waste Disposal Information

All toxic materials were disposed of in accordance with "Prudent Practices for Disposal of Chemicals from Laboratories"; National Academy Press; Washington, DC, 1983.

3. Discussion

The present procedure provides an example of the regiospecific synthesis of β–keto esters by the C-acylation of preformed lithium enolates with cyanoformate esters.[3] The enolates can be generated in a variety of ways, including direct enolization of ketones with suitable bases, liberation from silyl enol ethers and acetates, conjugate additions of cuprates to α,β-unsaturated enones and by the reduction of enones by lithium in liquid ammonia, as in the example described above. The more traditional acylating agents (acyl halides and anhydrides) afford variable amounts of O-acylated products[4] and although the use of reaction temperatures of -100°C is reported to obviate this problem with some substrates,[5] the method does not appear to be general. The classical solution has been to use carbon dioxide followed by esterification,[6] but yields rarely exceed 50%, indicating, *inter alia,* that the formation of enol carbonates probably occurs to a significant extent and that these unstable intermediates decompose to ketones on work up.[7] There is also the problem of handling thermally unstable β–keto acids. Carbon disulfide[8] and carbon oxysulfide[9] have been used instead of carbon dioxide, and the products methylated in situ to form dithio and thio esters, respectively. These can subsequently be converted into methyl esters by mercury(II) diacetate-catalyzed transesterification in methanol, but the methods are elaborate and do not appear to have found general use. Diethyl dicarbonate and its analogues show promise as possible alternatives to cyanoformates, but the reported yields to date are inferior.[10]

In the early investigations of this reaction it was found that in one comparative study of lithium, sodium and potassium enolates, only the lithium derivatives reacted satisfactorily,[3] although sodium enolates have subsequently been used successfully.[11] A range of cyanoformates can be employed, but the tert-butyl derivatives have been reported to be too unreactive. However, a satisfactory substitute which affords esters that can be de-alkylated with comparable facility is the p-methoxybenzyl analogue,[12] although it is prepared with some difficulty. In general, benzyl derivatives seem to give slightly better yields, possibly due to the more efficient isolation of products from substrates with lower molecular weights. Optically active cyanoformates derived from (+)-menthol, (-)-borneol, and the Oppolzer alcohol were reported to furnish good chemical yields, but the level of stereoselectivity was disappointingly low.[13]

For more conformationally-constrained chiral substrates, however, diastereoselectivity can be expected to be good to excellent. Lithium enolates derived from sterically unencumbered cyclohexanones undergo preferential axial acylation as illustrated by the reductive acylation of (R)-(-)-carvone **4** to afford a 3:1 mixture of esters **5** and **6**, whereas equatorial acylation is favored in compounds that possess an alkyl substituent in a 1,3-syn-axial relationship to the reacting center, as in the conversion of tricyclic enone **7** to ester **8** (epimeric with the product from the more traditional sequence of acylation followed by alkylation).[14] (In substrates of this kind it is assumed that the transition state structure is based on a twist-boat conformation which permits the reagent to approach along an axial-like trajectory on the less encumbered, lower face of the substrate.)[15]

[Scheme: compound 7 → 8 (>98% ds), reagents: 1. Li, NH₃, (CH₃)₃COH; 2. NCCO₂Me, Et₂O (71% yield)]

For compounds in which the β–carbon of the enolate is sterically hindered, this problem can normally be eliminated by the use of diethyl ether as the solvent. In several cases a switch from exclusive O-acylation in tetrahydrofuran to complete C-acylation in ether has been observed, and the conversion of **7** into **8** provides a typical illustration. When 3-methyl-2-cyclohexen-1-one **9** is converted into **10** by the addition of lithium dimethyl cuprate followed by in situ reaction with methyl cyanoformate, however, enol carbonate **11** accounts for 6% of the products. This can be avoided if the intermediate enolate is trapped as the enol trimethylsilyl ether and the enolate reliberated by treatment with butyllithium (note that the traditional reagent, methyllithium, is ineffective in ether as a solvent), but the overall yield is reduced.

[Scheme: compound 9 → 10 (78% yield) + 11 (6% yield), reagents: 1. Me₂CuLi, Et₂O, -5°C; 2. NCCO₂Me, -78°C]

Quite apart from the issue of regioselectivity, the present method is exceptionally reliable and makes it possible to prepare β-keto esters from ketones under especially mild conditions. It is not only the method of choice with sensitive substrates, but will often ensure superior results with more stable intermediates as well.[8,16] A wide range of examples is provided in the Table, while further examples have been reported elsewhere.[17] The methodology has also been applied to esters,[12] lactones,[18] and the N-acylation of lactams.[19] The formation of β–diketones by the treatment of lithium enolates with alkanoyl and aroyl nitriles has also been described.[20]

1. Research School of Chemistry, Australian National University, G.P.O. Box 4, Canberra, A.C.T., 2601, Australia.
2. Childs, M. E.; Weber, W. P. *J. Org. Chem.* **1976**, *41*, 3486.
3. Mander, L. N.; Sethi, S. P. *Tetrahedron Lett.* **1983**, *24*, 5425.
4. Caine, D. In "Carbon-Carbon Bond Formation"; Augustine, R. L., Ed.; Marcel Dekker: 1979; Vol. 1, pp 250-258.
5. Seebach, D.; Weller, T.; Protschuk, G.; Beck, A. K.; Hoekstra, M. S. *Helv. Chim. Acta* **1981**, *64*, 716.
6. Caine, D. *Org. React.* **1976**, *23*, 1-258.
7. Caine, D. In "Carbon-Carbon Bond Formation"; Augustine, R. L., Ed.; Marcel Dekker: 1979; Vol. 1, p 258, footnote 69.
8. Kende, A. S.; Becker, D. A. *Synth. Commun.* **1982**, *12*, 829.
9. Vedejs, E.; Nader, B. *J. Org. Chem.* **1982**, *47*, 3193.
10. Hellou, J.; Kingston, J. F.; Fallis, A. G. *Synthesis* **1984**, 1014.
11. Ziegler, F. E.; Klein, S. I.; Pati, U. K.; Wang, T.-F. *J. Am. Chem. Soc.* **1985**, *107*, 2730.
12. (a) Winkler, J. D.; Henegar, K. E.; Williard, P. G. *J. Am. Chem. Soc.* **1987**, *109*, 2850; (b) Henegar, K. E.; Winkler, J. D. *Tetrahedron Lett.* **1987**, *28*, 1051.
13. Kunisch, F.; Hobert, K.; Welzel, P. *Tetrahedron Lett.* **1985**, *26*, 5433.
14. Crabtree, S. R.; Chu, W. L. A.; Mander, L. N. *Synlett.* **1990**, 169
15. Evans, D. A. In "Asymmetric Synthesis"; Morrison, J. D., Ed.; Academic Press: Orlando, FL, 1984; Vol. 3, pp 1-110.
16. Mori, K.; Ikunaka, M. *Tetrahedron*, **1987**, *43*, 45.

17. (a) Miyata, O.; Hirata, Y.; Naito, T.; Ninomiya, I. *Heterocycles* **1984**, *22*, 1041; (b) Schuda, P. F.; Phillips, J. L.; Morgan, T. M. *J. Org. Chem.* **1986**, *51*, 2742; (c) Haynes, R. K.; Katsifis, A. G. *J. Chem. Soc., Chem. Commun.* **1987**, 340.
18. Ziegler, F. E.; Cain, W. T.; Kneisley, A.; Stirchak, E. P.; Wester, R. T. *J. Am. Chem. Soc.* **1988**, *110*, 5442.
19. (a) Melching, K. H.; Hiemstra, H.; Klaver, W. J.; Speckamp, W. N. *Tetrahedron Lett.* **1986**, *27*, 4799; (b) Esch, P. M.; Hiemstra, H.; Klaver, W. J.; Speckamp, W. N. *Heterocycles,* **1987,** *26,* 75.
20. Howard, A. S.; Meerholz, C. A.; Michael, J. P. *Tetrahedron Lett.* **1979**, 1339.

Appendix
Chemical Abstracts Nomenclature (Collective Index Number); (Registry Number)

Methyl cyanoformate: Formic acid, cyano-, methyl ester (8); Carbonocyanidic acid, methyl ester (9); (17640-15-2)

$\Delta^{1(9)}$-Octalone-2: 2(3H)-Naphthalenone, 4,4a,5,6,7,8-hexahydro- (8,9); (1196-55-0)

Isoprene (8); 1,3-Butadiene, 2-methyl- (9); (78-79-5)

TABLE
PREPARATION OF β-KETO-ESTERS BY THE KINETICALLY CONTROLLED C-ACYLATION OF LITHIUM ENOLATES WITH METHYL CYANOFORMATE

Entry	Substrate	Procedure[‡]	Product	Yield(%)
1	$CH_3(CH_2)_n$—C(=O)—CH₃	A	$CH_3(CH_2)_n$—C(=O)—CH₂—CO_2Me, n=1; n=4	84; 86
2	2-R-cyclohexanone	A	2-R-6-carbomethoxycyclohexanone, R=H; R=Me	86; 86
3	cycloheptanone	A	2-carbomethoxycycloheptanone	96
4	norbornan-2-one	A	3-carbomethoxynorbornan-2-one	85
5	2-R-cyclopentanone	A	2-R-5-carbomethoxycyclopentanone, R=H; R=Me	71; 75
6	1-OSiMe₃-2-methyl-cyclopentene/hexene (CH₂)ₙ	B	1-OSiMe₃-2-carbomethoxymethyl-cyclopentene/hexene (CH₂)ₙ, n=1; n=2	72; 68
7	(E)-hex-4-en-3-one	A	methyl (E)-3-oxo-hex-4-enoate	65
8	cyclohex-2-enone	A	6-carbomethoxycyclohex-2-enone	84
9	α-tetralone	A	2-carbomethoxy-α-tetralone	92
10	β-tetralone	A	1-carbomethoxy-β-tetralone	92
11	4a-methyl-4,4a,5,6,7,8-hexahydronaphthalen-2(3H)-one	C	decalone β-keto ester	85

[‡]Procedures: **A**: Reference 3; **B**: Lithium enolates generated by addition of methyllithium in THF at -78°C; **C**: As described for the conversion of enone **1** into β-keto ester **3**.

OXIDATION OF SECONDARY AMINES TO NITRONES
6-METHYL-2,3,4,5-TETRAHYDROPYRIDINE N-OXIDE
(Pyridine, 2,3,4,5-tetrahydro-6-methyl-, 1-oxide)

Submitted by Shun-Ichi Murahashi, Tatsuki Shiota, and Yasushi Imada.[1]
Checked by Gary C. Look and Larry E. Overman.

1. Procedure

In a 500-mL, three-necked, round-bottomed flask equipped with a 100-mL pressure-equalizing dropping funnel, a thermometer, and a magnetic stirring bar is placed 2.64 g (8.00 mmol) of sodium tungstate dihydrate (Note 1). After the flask is flushed with nitrogen, 40 mL of water and 23.5 mL (200 mmol) of 2-methylpiperidine (Note 2) are added. The flask is cooled with an ice-salt bath to -5°C (internal temperature) and 45.0 mL (440 mmol) of 30% aqueous hydrogen peroxide solution (Note 3) is added dropwise over a period of ca. 30 min. During the period of addition the reaction mixture should be carefully kept at a temperature below 20°C (Note 4). The cooling bath is removed, and the mixture is stirred for 3 hr (Note 5). Excess hydrogen peroxide is decomposed by adding ca. 3 g of sodium hydrogen sulfite with ice cooling (Note 6). The solution is saturated by adding ca. 25 g of sodium chloride and extracted with ten 200-mL portions of dichloromethane (Note 7). Combined organic extracts are dried over anhydrous sodium sulfate. The drying agent is removed by filtration, and the solvent is removed by a rotary evaporator keeping the temperature at 40°C (Note 8) to give a pale yellow oil (20.0-22.0 g), which may be sufficiently pure for some applications (Note 9). Purification of the nitrone is achieved by column chromatography on 300 g of silica gel packed in 97:3 chloroform/methanol in a 4.8-cm x 70-cm column (Note 10). The product is applied to the column in 10 mL of chloroform and the column is eluted with 97:3 chloroform/methanol. After twenty 100-mL fractions are collected, the eluent is changed to 8:2 chloroform/methanol, and another ten 100-mL fractions are collected and analyzed by thin layer chromatography (Note 11). Combination of fractions 16-30 and evaporation provides 14.0-15.7 g (62-70%) of pure 6-methyl-2,3,4,5-tetrahydropyridine N-oxide as a pale yellow oil (Notes 12 and 13).

2. Notes

1. Sodium tungstate dihydrate was purchased from Wako Pure Chemical Ind., Ltd. and used without further purification. The checkers employed material purchased from Mallinckrodt, Inc.

2. 2-Methylpiperidine purchased from Nacalai Tesque, Inc. was distilled prior to use (bp 119-120°C). The checkers employed 2-methylpiperidine purchased from Aldrich Chemical Company, Inc.

3. The 30% aqueous solution of hydrogen peroxide was purchased from Mitsubishi Gas Chemical Company, Inc. or Fisher Scientific. Ten percent excess of hydrogen peroxide is used to complete the reaction within an appropriate time.

4. This is an exothermic reaction. Higher reaction temperatures cause partial decomposition of the product.

5. The reaction mixture consists of the desired nitrone and 6-15% of isomeric 2-methyl-2,3,4,5-tetrahydropyridine N-oxide: ^{1}H NMR (CDCl$_3$, 500 MHz) δ: 1.53 (d, 3 H, J = 6.9, -CH$_3$), 7.14 (t, 1 H, J = 3.9, -CH=N-).

6. The presence of hydrogen peroxide is detected with potassium iodide-starch test paper.

7. Extraction with five 200-mL portions of dichloromethane gives 20-21 g of the product. Oxidation of secondary amines which have low molecular weights requires water as solvent. The nitrones thus obtained are highly soluble in water, and many extractions are required. However, other nitrones can be isolated easily by simple extraction.

8. Higher temperatures cause decomposition of the desired product, and lower temperatures retard the decomposition of the undesired nitrone to give the dimeric compound.

9. The crude nitrone consists of the desired nitrone (85-70%), the 1:1 adduct of the less substituted nitrone with the desired nitrone [(3,14-dimethyl-2,9-dioxa-1,8-diazatricyclo[8.4.0.0^{3,8}]tetradecane) (15-30%), R_f = 0.39 (TLC glass plate silica gel 60 F_{254}, obtained from E. Merck, 9:1 chloroform/methanol); m/e = 226.1681 ($C_{12}H_{22}N_2O_2$)], and the dimer of the desired nitrone [(3,10-dimethyl-2,9-dioxa-1,8-diazatricyclo[8.4.0.0^{3,8}]tetradecane) (< 1%), mp 87.5-88.0°C; R_f = 0.46 (under the same conditions); m/e = 226.1664]. The checkers found that the crude product decomposed noticeably when stored overnight at -20°C.

10. Silica gel 60 (70-230 mesh) was purchased from E. Merck. The checkers employed flash chromatography using a 20-cm x 7-cm column and 230-400 mesh EM silica gel 60. With this silica gel it is essential to have 1% triethylamine in the eluent.

11. The R_f value of the nitrone is 0.37 (under the same conditions described above).

12. The product has the following spectral characteristics: IR (neat) cm^{-1}: 2945, 1627, 1448, 1190, 1165, 951, 872, 750, a strong OH stretch at 3400 cm^{-1} is also apparent; ^{1}H NMR (CDCl$_3$, 500 MHz) δ: 1.71-1.77 (m, 2 H, H-4), 1.92-1.97 (m, 2 H, H-3), 2.11 (overlapping tt, 3 H, J = 1.5, 1.0, CH$_3$), 2.42-2.47 (m, 2 H, H-5), 3.78-3.83 (m, 2

H, H-2); ^{13}C NMR (CDCl$_3$, 68 MHz) δ: 18.0 (CH$_3$), 18.2, 22.7, 30.0 (C-5), 57.3 (C-2), 145.1 (C-6); UV (EtOH) 235 nm (ε 6910).

13. The nitrone slowly dimerizes at room temperature. It should be stored as a solution in a solvent such as dichloromethane to prevent dimerization.

Waste Disposal Information

All toxic materials were disposed of in accordance with "Prudent Practices for Disposal of Chemicals from Laboratories"; National Academy Press; Washington, DC, 1983.

3. Discussion

Nitrones are highly versatile synthetic intermediates and excellent spin trapping reagents.[2] In particular, nitrones are excellent 1,3-dipoles[3] and have been used for the synthesis of various nitrogen-containing biologically active compounds.[3a,3b] The preparation of nitrones has been performed either by condensation of aldehydes or ketones with hydroxylamines,[4] or by oxidation of the corresponding hydroxylamines.[5] The difficulty of these methods is in the preparation of the starting hydroxylamines. For example, cyclic hydroxylamines are prepared from the corresponding cyclic amines via thermal decomposition of the corresponding tertiary amine N-oxides.[6]

The present procedure provides a single step synthesis of nitrones from secondary amines.[7] Typical results of the preparation of nitrones are summarized in Table I. If necessary, the nitrones are easily purified by distillation, recrystallization, or column chromatography. Selenium dioxide is also an effective catalyst for the oxidation of secondary amines with hydrogen peroxide to give nitrones.[8] 1,3-Dipolar cycloadducts are obtained directly by the oxidation of secondary amines in the presence of alkenes.

The reaction of nitrones with various nucleophiles provides a powerful strategy for the introduction of a substituent at the α-position of secondary amines.[9] The reaction of nitrones with Grignard reagents or organolithium compounds affords various α-substituted hydroxylamines, which can be converted into α-substituted secondary amines by catalytic hydrogenation. The nucleophilic reaction with potassium cyanide gives α-cyanohydroxylamines which are useful precursors for amino acids and N-hydroxyamino acids.[10]

1. Department of Chemistry, Faculty of Engineering Science, Osaka University, Machikaneyama, Toyonaka, Osaka 560, Japan.
2. For reviews of nitrone chemistry, see: (a) Breuer, E. In "The Chemistry of Amino, Nitroso and Nitro Compounds and Their Derivatives"; Patai, S., Ed., Wiley, 1982; Part 1, pp. 459-564; (b) Tennant, G. In "Comprehensive Organic Chemistry"; Barton, D. H. R.; Ollis, W. D., Eds.; Pergamon Press, 1979; Vol. 2, pp. 500-510; (c) Hamer, J.; Macaluso, A. *Chem. Rev.* **1964**, *64*, 473-495.
3. (a) Tufariello, J. J. In "1,3-Dipolar Cycloaddition Chemistry"; Padwa, A., Ed.; Wiley, 1984; Vol. 2, pp. 83-168; (b) Tufariello, J. J. *Acc. Chem. Res.* **1979**, *12*, 396-403; (c) Black, D. St. C.; Crozier, R. F.; Davis, V. C. *Synthesis* **1975**, 205-221.
4. Sandler, S. R.; Karo, W. "Organic Functional Group Preparations"; Academic Press, 1983; Vol. 3, pp. 351-377.
5. Murahashi, S.-I.; Mitsui, H.; Watanabe, T.; Zenki, S.-i. *Tetrahedron Lett.* **1983**, *24*, 1049-1052, and references cited therein.
6. (a) Thesing, J.; Sirrenberg, W. *Chem. Ber.* **1959**, *92*, 1748-1755; (b) Thesing, J.; Mayer, H. *Justus Liebigs Ann. Chem.* **1957**, *609*, 46-57.

7. Murahashi, S.-I.; Mitsui, H.; Shiota, T.; Tsuda, T.; Watanabe, S. *J. Org Chem.* **1990**, *55*, 1736-1744.
8. Murahashi, S.-I.; Shiota, T. *Tetrahedron Lett.* **1987**, *28*, 2383-2386.
9. (a) Meyers, A. I. *Aldrichimica Acta* **1985**, *18*, 59-68; (b) Seebach, D.; Enders, D. *Angew. Chem., Inter. Ed. Engl.* **1975**, *14*, 15-32.
10. Murahashi, S.-I.; Shiota, T. *Tetrahedron Lett.* **1987**, *28*, 6469-6472.

Appendix
Chemical Abstracts Nomenclature (Collective Index Number); (Registry Number)

6-Methyl-2,3,4,5-tetrahydropyridine N-oxide: Pyridine, 2,3,4,5-tetrahydro-6-methyl-, 1-oxide (9); (55386-67-9)

Sodium tungstate dihydrate: Tungstic acid, disodium salt, dihydrate (8,9); (10213-10-2)

2-Methylpiperidine: Piperidine, 2-methyl- (8,9); (109-05-7)

Hydrogen peroxide (8,9); (7722-84-1)

TABLE
CATALYTIC OXIDATION OF SECONDARY AMINES WITH HYDROGEN PEROXIDE

Amine	Solvent	Product	Yield (%)
Bu-NH-Bu (di-n-butylamine)	CH_3OH	Bu-N$^+$(O$^-$)=CHCH$_2$CH$_2$CH$_3$ (nitrone)	89
iPr-NH-iPr	CH_3OH	iPr-N$^+$(O$^-$)=C(CH$_3$)$_2$	74
PhCH$_2$-NH-CH$_2$Ph	CH_3OH	PhCH$_2$-N$^+$(O$^-$)=CHPh	85
1,2,3,4-tetrahydroisoquinoline	CH_3OH	3,4-dihydroisoquinoline N-oxide	85
6-PhCH$_2$O-7-CH$_3$O-1,2,3,4-tetrahydroisoquinoline	CH_3OH	corresponding N-oxide	86
6-CH$_3$O-7-PhCH$_2$O-1-(3'-PhCH$_2$O-4'-CH$_3$O-benzyl)-1,2,3,4-tetrahydroisoquinoline	CH_3OH	corresponding N-oxide	60
6,7-methylenedioxy-1,2,3,4-tetrahydroisoquinoline	CH_3OH	corresponding N-oxide	62
pyrrolidine	H_2O	1-pyrroline N-oxide	44
piperidine	H_2O	2,3,4,5-tetrahydropyridine N-oxide	40

1-PHENYL-2,3,4,5-TETRAMETHYLPHOSPHOLE

(1H-Phosphole, 2,3,4,5-tetramethyl-l-phenyl-)

$(\eta\text{-}C_5H_5)_2ZrCl_2 \xrightarrow{\text{2-Butyne, BuLi}} (\eta\text{-}C_5H_5)_2Zr\overset{\displaystyle\diagup}{\underset{\displaystyle\diagdown}{}}$

$(\eta\text{-}C_5H_5)_2Zr\overset{\displaystyle\diagup}{\underset{\displaystyle\diagdown}{}} \xrightarrow{\text{PhPCl}_2} \text{PhP}\overset{\displaystyle\diagup}{\underset{\displaystyle\diagdown}{}}$

Submitted by Paul J. Fagan and William A. Nugent.[1]
Checked by Mark S. Jensen and James D. White.

1. Procedure

A 500-mL, three-necked, round-bottomed flask equipped with a magnetic stirring bar, rubber septum-capped pressure-equalizing addition funnel on the center neck, rubber septum on one side neck, and a nitrogen inlet with stopcock on the other side neck is charged in a nitrogen-filled glove box (Note 1) with 27.0 g (92.5 mmol) of zirconocene dichloride [$(\eta\text{-}C_5H_5)_2ZrCl_2$] (Note 2), 150 mL of tetrahydrofuran (Note 3), and 16.0 mL (204 mmol) of 2-butyne (Note 4) added via syringe. The apparatus is removed from the glove box and attached via the nitrogen stopcock to a nitrogen bubbler. The flask is cooled to -78°C (dry ice-acetone bath) and 108 mL of 1.72 M butyllithium (186 mmol) in hexane (Note 5) is added via syringe to the addition funnel through the septum. The butyllithium solution is added dropwise to the stirred mixture in the flask. After the addition is complete, the reaction mixture is stirred at -78°C for 10 min (Note 6). The flask is allowed to warm by removing the dry ice-acetone bath, and the reaction mixture is stirred at room temperature for 2.5 hr at which point the mixture is dark orange-red (Note 7). The flask is again cooled to -78°C, and 17.5 mL (129 mmol) of dichlorophenylphosphine (Note 8) is added in a slow stream via syringe through the septum on the flask. The dry ice-acetone bath is removed and the flask is allowed to warm to room temperature. After 1 hr the orange-red color has dissipated and the septa are replaced with glass stoppers. The nitrogen inlet is connected to a vacuum line (0.1 mm) and the solvent is removed under reduced pressure. The stopcock is closed and the apparatus is placed in a nitrogen-filled glove box (Note 1). The reaction residue is extracted three times with 30-mL portions of hexane, each of which is filtered from the precipitate of $(\eta\text{-}C_5H_5)_2ZrCl_2$ and combined. The hexane is removed under reduced pressure (0.5 mm), and the oily liquid that remains is transferred into a 100-mL round-bottomed flask. The flask is attached to a vacuum distillation apparatus with a 10-cm Vigreux column. The distillation apparatus is removed from the glove box and attached to a vacuum line. Distillation at 0.35 mm yields a fraction which is collected between 40-92°C and discarded. A second fraction boiling between 92-104°C (Note 9) is collected; the clear oily liquid is 1-phenyl-2,3,4,5-tetramethylphosphole [14.9-15.5 g, 75-78% based on $(\eta\text{-}C_5H_5)_2ZrCl_2)$] (Note 10). The compound is air-sensitive and should be stored under nitrogen.

2. Notes

1. A nitrogen-filled glove bag may also be used.
2. Zirconocene dichloride was obtained from Aldrich Chemical Company, Inc. and used without further purification.
3. Tetrahydrofuran was distilled from sodium benzophenone ketyl before use.
4. 2-Butyne was obtained from Farchan Chemical Company and dried over 4 Å molecular sieves before use.

5. Butyllithium was obtained from Foote Mineral Co. The molarity was checked by titration of 2.00 mL of the butyllithium solution in 10 mL of diethyl ether with dry 2-butanol using 1,10-phenanthroline as indicator.

6. The reaction mixture may become thick with a white solid at this point. The solid can be loosened by manually shaking the flask, or with the aid of an external permanent magnet using the magnetic stirring bar to break up the solid. The solid loosens some more upon warming, and stirring is not a problem.

7. If desired, the zirconium metallacycle can be isolated at this point: The flask is attached to a vacuum line via the nitrogen inlet, and solvents are removed from the flask under reduced pressure. With the aid of a 40°C water bath, the reaction residue is thoroughly dried. The flask is sealed and placed in a nitrogen-filled glove box (Note 1). The residue is extracted with small portions of toluene (total of 50 mL), each portion being filtered and combined. Toluene is removed from the filtrate under reduced pressure. Hexane (20 mL) is added to the solid residue, and after trituration, the solid is collected by filtration, and washed once with 10 mL of hexane. It is dried under reduced pressure to yield 26.0 g (85%) of crystalline, orange-red (η–$C_5H_5)_2ZrC_4(CH_3)_4$ which is >95% pure by spectroscopic analysis. This compound is very stable thermally both in solution and in the solid state; however, it is air-sensitive and should be handled under nitrogen. It can be used as obtained as a reagent in the synthesis of other heterocycles. The NMR spectrum is as follows: ^{1}H NMR (300 MHz, THF-d_8) δ: 1.54 (s, 6 H, CH_3), 1.57 (s, 6 H, CH_3), 6.15 (s, 10 H, η-C_5H_5).

8. Dichlorophenylphosphine (Strem Chemicals, Inc.) was vacuum distilled and placed under a nitrogen atmosphere before use.

9. The boiling point has been reported previously as 105-110°C at 0.5 mm.[2]

10. The product has the following spectral properties: ^{1}H NMR (300 MHz, THF-d_8) δ: 1.89 (d, 6 H, J = 10.6, CH_3), 1.93 (s, 6 H, CH_3), 7.24 (m, 5 H, phenyl); ^{13}C NMR (75.5 MHz, THF-d_8) δ: 12.9 (dq, J_{CH} = 126, J_{PC} = 22, CH_3), 13.9 (q, J_{CH} = 126, CH_3),

129.2 (dt, J_{CH} = 160 and 6, J_{PC} = 8, phenyl), 129.6 (dt, J_{CH} = 162 and 6, phenyl), 134.3 (ddt, J_{CH} = 161 and 5, J_{PC} = 20, phenyl), 134.7 (d, J_{PC} = 15), 136.4 (s, C=C), 143.4 (m, J_{PC} = 11); ^{31}P {^{1}H} NMR (121.7 MHz, THF-d$_8$) δ: 14 (s). Exact mass: Calcd. for C$_{14}$H$_{17}$P: m/e = 216.1068. Found: 216.1127.

Waste Disposal Information

All toxic materials were disposed of in accordance with "Prudent Practices for Disposal of Chemicals from Laboratories"; National Academy Press; Washington, DC, 1983.

3. Discussion

Phospholes and other related heterocycles are an important class of main group compounds. The chemistry of phospholes and their preparation has been reviewed extensively by Mathey.[3] We provide details here for a simple, one-pot procedure for the preparation of 1-phenyl-2,3,4,5-tetramethylphosphole applying zirconocene chemistry.[4] The procedure involves reduction of (η-C$_5$H$_5$)$_2$ZrCl$_2$ with butyllithium in the presence of 2-butyne which (as reported initially by Negishi, et al.[5]) forms a zirconium metallacycle. Addition of dichlorophenylphosphine to this reaction mixture produces the phosphole. One other procedure for the preparation of 1-phenyl-2,3,4,5-tetramethylphosphole has been reported by Nief, et al.[2] That procedure involved aluminum chloride - coupling of 2-butyne, followed by reaction with dichlorophenylphosphine to form a chlorophospholium tetrachloroaluminate which was then reduced with tributylphosphine to produce the phosphole in 68% yield.

Using a procedure similar to that described here, or using isolated zirconium metallacycles as reagents, we have been able to prepare not only phospholes, but also arsoles, stiboles, bismoles, siloles, germoles, stannoles, galloles, thiophenes, selenophenes, and borole Diels-Alder dimers.[4] Since a number of other titanium and zirconium metallacycles are readily available,[4] these reagents should be useful in the preparation of a variety of heterocycles.

1. E. I. du Pont de Nemours and Co., Inc., Central Research and Development Department, Experimental Station, Box 80328, E328/364, Wilmington, DE 19880-0328. Contribution No. 5123. We thank Ronald J. Davis for technical assistance.
2. Nief, F.; Mathey, F.; Ricard, L.; Robert, F. *Organometallics* **1988**, *7*, 921.
3. Mathey, F. *Chem. Rev.* **1988**, *88*, 429; Mathey, F. *Phosphorus Sulfur* **1983**, *18*, 101; Mathey, F.; Fischer, J.; Nelson, J. H. *Struct. Bonding (Berlin)* **1983**, *55*, 153; Mathey, F. In "Topics in Phosphorus Chemistry;" Grayson, M.; Griffith, E. J., Eds.; Wiley: New York, 1980; Vol. 10, p. 1-128.
4. Fagan, P. J.; Nugent, W. A. *J. Am. Chem. Soc.* **1988**, *110*, 2310; Fagan, P. J.; Burns, E. G.; Calabrese, J. C. *J. Am. Chem. Soc.* **1988**, 110, 2979; RajanBabu, T. V.; Nugent, W. A.; Taber, D. F.; Fagan, Paul, J. *J. Am. Chem. Soc.* **1988**, *110*, 7128; Buchwald, S. L.; Watson, B. T.; Lum, R. T.; Nugent, W. A. *J. Am. Chem. Soc.* **1987**, *109*, 7137.
5. Negishi, E.-i.; Cederbaum, F. E.; Takahashi, T. *Tetrahedron Lett.* **1986**, *27*, 2829.

Appendix

Chemical Abstracts Nomenclature (Collective Index Number); (Registry Number)

1-Phenyl-2,3,4,5-tetramethylphosphole: 1H-Phosphole, 2,3,4,5-tetramethyl-1-phenyl- (12); (112549-07-2)

Zirconocene dichloride: Zirconium, dichloro-π-cyclopentadienyl- (8); Zirconium, dichlorobis(η^5-2,4-cyclopentadien-1-yl)- (9); (1291-32-3)

2-Butyne (8,9); (503-17-3)

Dichlorophenylphosphine: Phosphonous dichloride, phenyl- (8,9); (644-97-3)

Unchecked Procedures

Accepted for checking during the period June 1, 1990 through May 1, 1991. An asterisk (*) indicates that the procedure has been subsequently checked.

In accordance with a policy adopted by the Board of Editors, beginning with Volume 50 and further modified subsequently, procedures received by the Secretary and subsequently accepted for checking will be made available upon request to the Secretary, if the request is accompanied by a stamped, self-addressed envelope. (Most manuscripts require 54¢ postage).

Address requests to:

Professor Jeremiah P. Freeman
Organic Syntheses, Inc.
Department of Chemistry
University of Notre Dame
Notre Dame, Indiana 46556

It should be emphasized that the procedures which are being made available are unedited and have been reproduced just as they were first received from the submitters. There is no assurance that the procedures listed here will ultimately check in the form available, and some of them may be rejected for publication in *Organic Syntheses* during or after the checking process. For this reason, *Organic Syntheses* can provide no assurance whatsoever that the procedures will work as described and offers no comment as to what safety hazards may be involved. Consequently, more than usual caution should be employed in following the directions in the procedures.

Organic Syntheses welcomes, on a strictly voluntary basis, comments from persons who attempt to carry out the procedures. For this purpose, a Checker's Report form will be mailed out with each unchecked procedure ordered. Procedures which have been checked by or under the supervision of a member of the Board of Editors will continue to be published in the volumes of *Organic Syntheses*, as in the past. It is anticipated that many of the procedures in the list will be published (often in revised form) in *Organic Syntheses* in future volumes.

2551R*	Tributyl (3-methyl-2-butenyl)tin. Y. Naruta, Y. Nishigaichi, and K. Maruyama, Department of Chemistry, Faculty of Science, Kyoto University, Sakyo-ku, Kyoto 606, Japan
2552R*	Ubiquinone-1. Y. Naruta and K. Maruyama, Department of Chemistry, Faculty of Science, Kyoto University, Sakyo-ku, Kyoto 606, Japan
2571*	9-Bromo-9-phenylfluorene. T. F. Jamison, W. D. Lubell, J. M. Dener, M. J. Krisché and H. Rapoport, Department of Chemistry, University of California, Berkeley, CA 94720
2576*	Benzoannelation of Ketones. M. A. Tius and G. S. K. Kannangara, Department of Chemistry, The University of Hawaii, Honolulu, Hawaii 96822
2577*	3-(1-Octen-1-yl)cyclopentanone. R. C. Sun, M. Okabe, D. L. Coffen, and J. Schwartz, Chemistry Research Department, Hoffmann-La Roche Inc., Nutley, NJ 07110
2578	Asymmetric Hydrogenation of Allylic Alcohols Using BINAP-Ruthenium Complexes: (S)-(-)-Citronellol. H. Takaya, T. Ohta, S.-i. Inoue, and R. Noyori, Department of Chemistry, Nagoya University, Chikusa, Nagoya 464, Japan
2579R*	Asymmetric Hydrogenation of 3-Oxo Carboxylates Using BINAP-Ruthenium Complexes: (R)-(-)-Methyl 3-Hydroxybutanoate. M. Kitamura, N. Sayo, and R. Noyori, Department of Chemistry, Nagoya University, Chikusa, Nagoya 464, Japan
2581*	Schwartz's Reagent. S. L. Buchwald, S. J. LaMarie, R. B. Nielsen, B. T. Watson, and S. M. King, Department of Chemistry, Massachusetts Institute of Technology, Cambridge, MA 02139
2583*	An Efficient Synthesis of Indole-2-acetic Acid Methyl Esters. S. P. Modi, R. C. Oglesby, and S. Archer, Department of Chemistry, Rensselaer Polytechnic Institute, Troy, NY 12180
2586*	Direct Large-Scale-Degradation of the Biopolymer Polyhydroxybutanoate to (R)-3-Hydroxybutanoic Acid and its Methyl Ester. D. Seebach, A. K. Beck, R. Breitschuh and K. Job, Laboratorium für Organische Chemie der Eidgenössischen Technischen Hochschule, ETH-Zentrum, Universitätstrasse 16, CH-8092 Zürich, Switzerland

2588*	Palladium(O)-Catalyzed Reaction of 9-Alkyl-9-borabicyclo[3.3.1]-nonane with 1-Bromo-1-phenylthioethene: 4-(3-Cyclohexenyl)-2-phenylthio-1-butene. T. Ishiyama, N. Miyaura, and A. Suzuki, Department of Applied Chemistry, Faculty of Engineering, Hokkaido University, Sapporo 060, Japan
2589	(R)-Tetrahydro-1-methyl-3,3-diphenyl-1H,3H-pyrrolo[1,2,c][1,3,2]oxazaborole. L. C. Xavier and J. J. Mohan, Merck Sharp & Dohme Research Laboratories, P.O. Box 2000, Rahway, NJ 07065
2590*	Preparation and Use of (Methoxymethoxy)-methyllithium: 1-(Hydroxymethyl)-cycloheptanol. C. R. Johnson, J. R. Medich, R. L. Danheiser, K. R. Romines, H. Koyama, and S. K. Gee, Department of Chemistry, Wayne State University, Detroit, MI 48202
2591*	A Hydroxymethyl Anion Equivalent: Tributyl-[(methoxymethoxy)methyl]stannane. R. L. Danheiser, K. R. Romines, H.Koyama, S. K. Gee, C. R. Johnson, and J. R. Medich, Department of Chemistry, Wayne State University, Detroit, MI 48202
2592	Diastereoselective Homologation of (R)-2,3-O-Isopropylideneglyceraldehdye Using 2-(Trimethylsilyl)thiazole: 2-O-Benzyl-3,4-isopropylidene-D-erythrose. A. Dondoni, G. Fantin, and M. Fogagnolo, Dipartimento di Chimica, Laboratorio di Chimica Organica, Universitá, 44100 Ferrara, Italy
2593R	Spiroannelation of Enolsilanes: 2-Oxo-5-methoxyspiro[4.5]decane. T. V. Lee and J. R. Porter, School of Chemistry, University of Bristol, Cantock's Close, Bristol, England BS8 1TS
2595*	(R)-(+)-2-Hydroxy-1,2,2-triphenylethyl Acetate. M. Braun, S. Schneider, and S. Houben, Institut für Organische und Makromolekulare Chemie, Heinrich-Heine Universität, Universitätsstrasse 1, D-4000 Düsseldorf, West Germany
2597	Benzocyclobutenone by Flash Vacuum Pyrolysis. P. Schiess, P. V. Barve, F. E. Dussy, and A. Pfiffner, Institut für Organische Chemie, Universität Basel, St. Johannsring 19, CH-4056 Basel, Switzerland
2598*	(1R,2R)-(+)- and (1S,2S)-(-)-1,2-Diphenyl-1,2-ethylenediamine. S. Pikul and E. J. Corey, Department of Chemistry, Harvard University, Cambridge, MA 02138

2599R	Enantioselective, Catalytic Diels-Alder Reaction: (1S-endo)-3-(Bicyclo[2,2,1]hept-5-en-2-ylcarbonyl)-2-oxazolidinone. S. Pikul and E. J. Corey, Department of Chemistry, Harvard University, Cambridge, MA 02138
2602*	Diethyl 1-Propyl-2-oxoethylphosphonate. P. Savignac and C. Patois, Laboratoire de Chimie du Phosphore et des Métaux de Transtition, DCPH, Ecole Polytechnique, 91128 Palaiseau Cedex, France
2603*	Diastereoselective Formation of trans-1,2-Disubstituted Cyclohexanes from Alkylidenemalonates by an Intramolecular Ene Reaction: Dimethyl (1'R,2'R,5'R)-2-(2'-Isopropenyl-5'-methylcyclohex-1'-yl)-propane-1,3-dioate[4a]. L. F. Tietze and U. Beifuss, Institut für Organische Chemie der Universität Göttingen, Tammannstr. 2, D-3400 Göttingen, Federal Republic of Germany
2605*	2-Methylene-1,3-dithiolane. K. R. Dahnke and L. A. Paquette, Department of Chemistry, The Ohio State University, Columbus, OH 43210
2606R*	Inverse Electron Demand Diels-Alder Cycloaddition of a Ketene Dithioacetal. Copper Hydride-Promoted Reduction of a Conjugated Enone. 9-Dithiolanobicyclo[3.2.2]non-en-2-one from Tropone. K. R. Dahnke and L. A. Paquette, Department of Chemistry, The Ohio State University, Columbus, OH 43210
2607*	A Selective, Heterogeneous Oxidation Using a Mixture of Potassium Permanganate and Cupric Sulfate. C. W. Jefford, Y. Li, and Y. Wang, Department of Organic Chemistry, University of Geneva, 1211 Geneva 4, Switzerland
2610*	Ethyl 3-Oxo-4-pentenoate (Nazarov's Reagent). R. Zibuck and J. Streiber, Department of Chemistry, Wayne State University, Detroit, MI 48202
2611*	3(S)-((tert-Butyldiphenylsilyl)oxy)-2-butanone. L. E. Overman and G. M. Rishton, Department of Chemistry, University of California, Irvine, CA 92717
2612*	Stereocontrolled Preparation of 3-Acyltetrahydrofurans from Acid-Promoted Rearrangements of Allylic Ketals: (2S,3S)-3-Acetyl-8-carboethoxy-2,3-dimethyl-1-oxa-8-azaspiro-[4.5]decane. L. E. Overman and G. M. Rishton, Department of Chemistry, University of California, Irvine, CA 92717

CUMULATIVE AUTHOR INDEX
FOR VOLUME 70

This index comprises the names of contributors to Volume 70 only. For authors to previous volumes, see either indices in Collective Volumes I through VII or the single volume entitled *Organic Syntheses, Collective Volumes, I, II, III, IV, V, Cumulative Indices,* edited by R. L. Shriner and R. H. Shriner.

Arnold, H., **70**, 111
Arnold, L. D., **70**, 1, 10

Baum, J. S., **70**, 93
Bigham, M. H., **70**, 231
Bishop, R., **70**, 120
Boger, D. L., **70**, 79
Bonin, M., **70**, 54
Bradshaw, J. S., **70**, 129
Brown, D. S., **70**, 157

Cantrell, Jr., W. R., **70**, 93
Carrasco, M., **70**, 29
Cava, M. P., **70**, 151
Chen, H. G., **70**, 195
Chen, W., **70**, 151
Crabtree, S. R., **70**, 256

Davies, H. M. L., **70**, 93
Dickhaut, J., **70**, 164

Fagan, P. J., **70**, 272

Garner, P., **70**, 18
Gibson, F. S., **70**, 101
Giese, B., **70**, 164
Gilheany, D. G., **70**, 47
Grierson, D. S., **70**, 54

Hamper, B. C., **70**, 246
Helquist, P., **70**, 177
Husson, H.-P., **70**, 54
Huyer, G., **70**, 1

Imada, Y., **70**, 265

Jackson, Y. A., **70**, 151
Johnson, M. R., **70**, 68
Jones, R. J., **70**, 29

Kamel, S., **70**, 29
Kazlauskas, R. J., **70**, 60
Kitamura, T., **70**, 215
Knapp, S., **70**, 101
Knochel, P., **70**, 195
Krakowiak, K. E., **70**, 129

Lenz, G. R., **70**, 139
Lessor, R. A., **70**, 139
Ley, S. V., **70**, 157
Luh, T.-Y., **70**, 240

Mander, L. N., **70**, 256
Mattson, M. N., **70**, 177
McDonald, F. E., **70**, 204
McKee, B. H., **70**, 47
Meister, P. G., **70**, 226
Mozaffari, A., **70**, 68
Murahashi, S.-I., **70**, 265
Myles, D. C., **70**, 231

Negron, A., **70**, 169
Ni, Z.-J., **70**, 240
Nugent, W. A., **70**, 272

O'Connor, E. J., **70**, 177
Overman, L. E., **70**, 111

Panek, J. S., **70**, 79
Pansare, S. V., **70**, 1, 10
Paquette, L. A., **70**, 226
Park, J. M., **70**, 18
Patel, M., **70**, 79
Pfau, M., **70**, 35

Rapoport, H., **70**, 29
Revial, G., **70**, 35
Romines, K. R., **70**, 93
Royer, J., **70**, 54

Sessler, J. L., **70**, 68
Sethi, S. P., **70**, 256
Sharp, M. J., **70**, 111
Sharpless, K. B., **70**, 47
Shiota, T., **70**, 265
Sivik, M. R., **70**, 226
Soderquist, J. A., **70**, 169
Stang, P. J., **70**, 215
Stephenson, E. K., **70**, 151

Truong, T., **70**, 29

Vederas, J. C., **70**, 1, 10

Wender, P. A., **70**, 204
White, A. W., **70**, 204
Witschel, M. C., **70**, 111

Yeh, M. C. P., **70**, 195

CUMULATIVE SUBJECT INDEX
FOR VOLUME 70

This index comprises subject matter for Volume **70** only. For subjects in previous volumes, see the cumulative indices in Volume **69**, which covers Volumes **65** through **69** and either the indices in Collective Volumes I through VII or the single volume entitled *Organic Syntheses, Collective Volumes I, II, III, IV, V, Cumulative Indices*, edited by R. L. Shriner and R. H. Shriner.

The index lists the names of compounds in two forms. The first is the name used commonly in procedures. The second is the systematic name according to **Chemical Abstracts** nomenclature. Both are usually accompanied by registry numbers in parentheses. Also included are general terms for classes of compounds, types of reactions, special apparatus, and unfamiliar methods.

Most chemicals used in the procedure will appear in the index as written in the text. There generally will be entries for all starting materials, reagents, intermediates, important by-products, and final products. Entries in capital letters indicate compounds appearing in the title of the preparation.

Acetamide, 2-(2-chloroethoxy)-N-ethyl-; (36961-73-6), **70**, 130

Acetamide, N-[2-(2-hydroxyethoxy)ethyl]-; (118974-46-2), **70**, 130

Acetamidoacrylic acid, **70**, 6

p-Acetamidobenzenesulfonyl azide, **70**, 93

p-Acetamidobenzenesulfonyl chloride, **70**, 93

Acetic acid, bromo-, ethyl ester; (105-36-2), **70**, 248

Acetic acid, diazo-, ethyl ester; (623-73-4), **70**, 80

Acetic acid, isocyano-, ethyl ester; (2999-46-4), **70**, 70

Acetic acid, lead (4+) salt; (546-67-8), **70**, 140

Acetic acid, rhodium (2+) salt; (5503-41-3), **70**, 93

Acetic acid, trifluoro-; (76-05-1), **70**, 141

Acetic acid, trifluoro-, anhydride; (407-25-0), **70**, 246

Acetic anhydride; (108-24-7), **70**, 69, 130

Acetoacetic acid, 2-diazo-, ethyl ester, **70**, 94

Acetoacetic acid, ethyl ester; (141-97-9), **70**, 94

4-Acetoxy-3-nitrohexane, **70**, 69

Acetylenes, perfluoroalkyl, **70**, 246

α-ACETYLENIC ESTERS, **70**, 246

O-Acetylserine, **70**, 6

N-Acylaziridines, **70**, 106

α-Acylmethylenephosphoranes, **70**, 251

α-Alkylated piperidines, **70**, 57

Alkylzinc iodide, **70**, 201

ALKYNE-IMINIUM ION CYCLIZATIONS, **70**, 111

Alkynyl(phenyl)iodonium sulfonates, **70**, 223

ALKYNYL(PHENYL)IODONIUM TOSYLATES, **70**, 215

ALLYLIC DITHIOACETALS, **70**, 240

Aluminum chloride; (7446-70-0), **70**, 226

Aluminum hydride reducing agents, **70**, 65

Aluminum, hydrobis(2-methylpropyl)-; (1191-15-7), **70**, 20

Aluminum, hydrodiisobutyl-, **70**, 20

Aluminum trichloride, **70**, 162

α-Amino acids, **70**, 6

β-Amino alcohols, **70**, 58

2-(2-Aminoethoxy)ethanol, **70**, 129

Amino lactams, **70**, 106

α-Aminonitrile, **70**, 57

(S)-3-AMINO-2-OXETANONE P-TOLUENESULFONATE, **70**, 10

Amipurimycin studies, **70**, 27

Anhydrous acetonitrile, **70**, 2

Annulated lactams, **70**, 106

Arsoles, **70**, 276

Asymmetric dihydroxylation, **70**, 51

ASYMMETRIC SYNTHESIS OF SUBSTITUTED PIPERIDINES, **70**, 54

Azides, sulfonyl, **70**, 97

β-Azidoalanine, **70**, 15

Azido lactams, **70**, 106

Azobisisobutyronitrile, **70**, 151, 167

Barton-Zard pyrrole synthesis, **70**, 75

Benzazepine, **70**, 145

Benzazepinone, **70**, 144

2H-3-Benzazepin-2-one, 1,3,4,5-tetrahydro-7,8-dimethoxy-; (20925-64-8), **70**, 139

Benzene, (1-butyl-1-octen-3-ynyl)-, (E)-; (111525-79-2), **70**, 215

Benzene, 1,1'-(1,2-ethenediyl)bis-, (E)-; (103-30-0), **70**, 47

Benzene, 1,1'-ethylidenebis-; (530-48-3), **70**, 179

Benzene, ethynyl-; (536-74-3), **70**, 157

Benzene, iodoso-, **70**, 217

Benzenemethanamine, α-methyl-, (S)-; (2627-86-3), **70**, 35

Benzenemethanamine, α-methyl-N-(2-methylcyclohexylidene)-; (76947-33-6), **70**, 35

Benzenesulfonyl azide, 4-(acetylamino)-; (2158-14-7), **70**, 93

Benzenesulfonyl chloride, 4-(acetylamino)-; (121-60-8), **70**, 93

2-BENZENESULFONYL CYCLIC ETHERS, **70**, 157

Benzenethiol, copper(1+) salt; (1192-40-1), **70**, 206

Benzylamine; (100-46-9), **70**, 112, 131

Benzylamine, α-methyl-, (S)-(-)-, **70**, 35

Benzyl chloroformate, **70**, 29

4-BENZYL-10,19-DIETHYL-4,10,19-TRIAZA-1,7,13,16-TETRAOXA-
 CYCLOHENEICOSANE (TRIAZA-21-CROWN-7), **70**, 129, 132

N-Benzyl-4-hexyn-1-amine; (112069-91-7), **70**, 112

(E)-1-BENZYL-3-(1-IODOETHYLIDENE)PIPERIDINE, **70**, 111

N-(Benzyloxycarbonyl)-L-methionine methyl ester, **70**, 29

N$^\alpha$-(BENZYLOXYCARBONYL)-β-(PYRAZOL-1-YL)-L-ALANINE, **70**, 2

N-(BENZYLOXYCARBONYL)-L-SERINE, **70**, 1

N-(Benzyloxycarbonyl)-L-serine β-lactone, **70**, 1

N-(BENZYLOXYCARBONYL)-L-VINYLGLYCINE METHYL ESTER, **70**, 29

9-Benzyl-3,9,15-triaza-6,12-dioxaheptadecane, **70**, 130

Bicyclo[2.2.1]heptane-1-methanesulfonic acid, 7,7-dimethyl-2-oxo-, (±);
 (5872-08-2), **70**, 112

[1,1'-Binaphthalene]-2,2'-diol, (R)-(+)-; (18531-94-7), **70**, 60

[1,1'-Binaphthalene]-2,2'-diol, (S)-(-)-; (18531-99-2), **70**, 60

(R)-(+)-1,1'-BI-2-NAPHTHOL, **70**, 60

(S)-(-)-1,1'-BI-2-NAPHTHOL, **70**, 60

(±)-1,1'-Bi-2-naphthyl pentanoate, **70**, 61

(R)-1,1'-Bi-2-naphthyl pentanoate, **70**, 62

Bismoles, **70**, 276

1,2-Bis(2-iodoethoxy)ethane, **70**, 132

Bis(trimethylsilyl)imidate, **70**, 104

Bis(trimethylsilyl)methyllithium, **70**, 243

9-Borabicyclo[3.3.1]nonane; (280-64-8), **70**, 169

9-BORABICYCLO[3.3.1]NONANE DIMER; (70658-61-6), **70**, 169

9-Borabicyclo[3.3.1]nonane, 9-methoxy-; (38050-71-4), **70**, 173

Boran-tetrahydrofuran, **70**, 131

Borane, compd. with thiobis[methane] (1:1); (13292-87-0), **70**, 169

Borane-methyl sulfide complex, **70**, 169

Borate (1-), tetrafluoro-, sodium; (13755-29-8), **70**, 178

Borole, Diels-Alder dimers, **70**, 276

Boron trifluoride etherate, **70**, 22, 197, 217, 240

Bovine pancreas acetone powder, **70**, 61

1-Bromobutane; (109-65-9), **70**, 218

Bromocresol green spray, **70**, 5, 13

2-(3-Bromopropyl)-1H-isoindole-1,3(2H)-dione, **70**, 195

2-Bromopyrrole, **70**, 155

N-Bromosuccinimide, **70**, 155

Butane, 1,4-dichloro-; (110-56-5), **70**, 206

Butanoic acid, 2-diazo-3-oxo-, ethyl ester; (2009-97-4), **70**, 94

Butanoic acid, 4,4,4-trifluoro-3-oxo-2-(triphenylphosphoranylidene)-,
    ethyl ester; (83961-56-2), **70**, 246

2-Butanol, (±)-; (15892-23-6), **70**, 206

2-Butanone, 4-[1-methyl-2-[(1-phenylethyl)imino]cyclohexyl]-,
[S-(R*, S*)]-; (94089-44-8), **70**, 36

3-Butenoic acid, 2-amino-, hydrochloride, (S)-; (75266-38-5), **70**, 30

Butenoic acid, 4-(methylsulfinyl)-2-[[(phenylmethoxy)carbonyl]amino]-,
methyl ester, (S)-; (75266-39-3), **70**, 30

3-Butenoic acid, 2-[[(phenylmethoxy)carbonyl]amino]-, methyl ester,
(S)-; (75266-40-9), **70**, 29

3-Buten-2-one; (78-94-4), **70**, 36

N-tert-BUTOXYCARBONYL-2-BROMOPYRROLE, **70**, 151

N-(tert-Butoxycarbonyl)-6,7-dimethoxy-1-methylene-1,2,3,4-tetrahydroisoquinoline,
70, 139

N-tert-Butoxycarbonylserine, **70**, 19

N-tert-Butoxycarbonyl-L-serine, **70**, 11

N-tert-BUTOXYCARBONYL-L-SERINE β-LACTONE, **70**, 10

N-tert-BUTOXYCARBONYL-2-TRIMETHYLSILYLPYRROLE, **70**, 151

N-tert-Butoxy-2,5-disubstituted pyrroles, **70**, 155

Butyllithium, **70**, 152, 154, 272

2-Butyne; (503-17-3), **70**, 272

2-Butynoic acid, 4,4,4-trifluoro-, ethyl ester; (79424-03-6), **70**, 246

Calicheamycin fragment, **70**, 27

Camphorsulfonic acid monohydrate, **70**, 112

Carbamic acid, (2-oxo-3-oxetanyl)-, 1,1-dimethylethyl ester, (S)-; (98541-64-1), **70**, 10

5-Carbamoylpolyoxamic acid, **70**, 27

(Carbethoxymethyl)triphenylphosphonium bromide, **70**, 246

p-Carboxybenzenesulfonyl azide, **70**, 97

Celite filter, **70**, 122

3-Chloro-5,5-dimethylcyclohex-2-en-1-one, **70**, 205

N-[2-(2-Chloroethoxy)ethyl]acetamide, **70**, 130

Chloromethyl methyl sulfide, **70**, 177

Chloromethyltrimethylsilane, **70**, 241

Chlorotrimethylsilane; (75-77-4), **70**, 152, 164, 196, 216

Cholesterol esterase, **70**, 61

Cinchonan-9-ol, 10,11-dihydro-6'-methoxy-, 4-chlorobenzoate (ester), (9S)-; (113162-02-0), **70**, 47

(E)-Cinnamaldehyde, **70**, 197, 240

Citric acid, **70**, 54

CONJUGATED ENYNES, **70**, 215

Copper(I) bromide-dimethyl sulfide complex, **70**, 218

Copper cyanide; (544-92-3), **70**, 196

Copper(I) thiophenoxide, **70**, 206

Copper-zinc organometallic, **70**, 196

Crown ethers, chiral, **70**, 65

2-CYANO-6-PHENYLOXAZOLOPIPERIDINE, **70**, 54

1,3-Cyclohexanedione, 5,5-dimethyl-; (126-81-8), **70**, 205

Cyclohexanone, 2-methyl; (583-60-8), **70**, 35

2-Cyclohexen-1-one, 3-chloro-5,5-dimethyl-; (17530-69-7), **70**, 205

Cyclohexylporphyrin, **70**, 77

1,5-Cyclooctadiene, (Z,Z)-; (1552-12-1), **70**, 120

1,5-Cyclooctadiene; (111-78-4), **70**, 169

Cyclopentadienyliron dicarbonyl dimer, **70**, 177

Cyclopentadienyliron dicarbonyl dimethsulfonium tetrafluoroborate; (72120-26-4), **70**, 179

1,3-Cyclopentanedione, 2-methyl-; (765-69-5), **70**, 226

Cyclopenta[b]pyrrol-2(1H)-one, hexahydro-6-iodo-, (3aα, 6α, 6aα)-; (100556-58-9), **70**, 101

2-Cyclopentene-1-acetamide; (72845-09-1), **70**, 101

2-Cyclopentene-1-acetic acid; (13668-61-6), **70**, 101

2-(2-Cyclopentenyl)acetyl chloride, **70**, 102

Cyclopropanation, **70**, 187

Cyclopropane, 1,1-diphenyl-; (3282-18-6), **70**, 177

Cyclopropenes, **70**, 97

DABCO, **70**, 253

Decahydroquinolines, **70**, 57

1,4-Diazabicyclo[2.2.2]octane, **70**, 253

1,8-Diazabicyclo[5.4.0]undec-7-ene, **70**, 70

Diazenedicarboxylic acid, diethyl ester; (1972-28-7), **70**, 10

Diazenedicarboxylic acid, dimethyl ester; (2446-84-6), **70**, 1

Diazomethane, **70**, 22

Diazo transfer reaction, **70**, 97

1,3-Dibromo-5,5-dimethylhydantoin, **70**, 151

1,2-Dibromoethane; (106-93-4), **70**, 196

Di-tert-butyl dicarbonate, **70**, 152

Di-tert-butyl pyrocarbonate, **70**, 140

Dicarbonic acid, bis(1,1-dimethylethyl) ester; (24424-99-5), **70**, 18, 140, 152

Dichlorobis(triphenylphosphine)nickel, **70**, 241

1,4-Dichlorobutane, **70**, 206

Dichlorophenylphosphine, **70**, 273

(1α,2α,5α,6α)-2,6-Dichloro-9-thiabicyclo[3.3.1]nonane; (10502-30-4), **70**, 120

2,6-Dicyanopiperidines, **70**, 55

DIELS-ALDER REACTION OF AN HETEROCYCLIC AZADIENE, **70**, 79

Diels-Alder reaction, inverse demand, **70**, 79

Diels-Alder reactions, **70**, 85

2-N,N'-Diethylaminomethyl-3,4-diethylpyrrole, **70**, 74

Diethyl azodicarboxylate, **70**, 4, 10

3,4-Diethyl-5-hydroxymethylpyrrole-2-carboxylic acid, **70**, 74

3,4-DIETHYLPYRROLE, **70**, 68

Dihalocarbene, **70**, 186

3,4-Dihydro-6,7-dimethoxy-1-methylene, 1,1-dimethylethyl ester; (82044-08-4), **70**, 139

Dihydroisoquinolines, **70**, 145

1,4-DIHYDROPYRIDINE EQUIVALENT, **70**, 54

Dihydroquinidine, benzoate ester, **70**, 49

Dihydroquinidine 4-chlorobenzoate, **70**, 47

Dihydroquinidine, 2-naphthoate ester, **70**, 49

Dihydro-1,2,4,5-tetrazine-3,6-dicarboxylate, **70**, 80

Diisobutylaluminum hydride, **70**, 20

1,4- and 1,5-Dilithioalkanes, **70**, 211

1,4-Dilithiobutane, **70**, 205

Dimedone, **70**, 205

Dimethoxyethane; (110-71-4), **70**, 169

6,7-Dimethoxy-1-methyl-3,4-dihydroisoquinoline, **70**, 139

2,2-Dimethoxypropane, **70**, 22

7,8-DIMETHOXY-1,3,4,5-TETRAHYDRO-2H-3-BENZAZEPIN-2-ONE, **70**, 139

Dimethyl azodicarboxylate, **70**, 1

Dimethyl dihydro-1,2,4,5-tetrazine-3,6-dicarboxylate, **70**, 80

N-[(1,1-Dimethylethoxy)carbonyl]-, methyl ester; (2766-43-0), **70**, 18

N-[(1,1-Dimethylethoxy)carbonyl]-L-serine methyl ester, **70**, 18

1,1-DIMETHYLETHYL (S)- OR (R)-4-FORMYL-2,2-DIMETHYL-3-OXAZOLIDINECARBOXYLATE, **70**, 18

3-(1,1-Dimethylethyl) 4-methyl (S)-2,2-dimethyl-3,4-oxazolidine-dicarboxylate, **70**, 19

2,2-Dimethyl-3,4-oxazolidinedicarboxylate, **70**, 19

Dimethyl 4-phenyl-1,2-diazine-3,6-dicarboxylate, **70**, 82

Dimethyl 3-phenylpyrrole-2,5-dicarboxylate, **70**, 82

9,9-DIMETHYLSPIRO[4.5]DECAN-7-ONE, **70**, 204

Dimethyl sulfide; (75-18-3), **70**, 218

Dimethyl sulfoxide; (67-68-5), **70**, 112, 122

DIMETHYL 1,2,4,5-TETRAZINE-3,6-DICARBOXYLATE, **70**, 79

2,6-Dinitrato-9-thiabicyclo[3.3.1]nonane 9,9-dioxide, **70**, 125

1,4-Dioxane; (123-91-1), **70**, 84, 179

1,1-DIPHENYLCYCLOPROPANE, **70**, 177

(R,R)-1,2-DIPHENYL-1,2-ETHANEDIOL, **70**, 47

(E)-1,2-Diphenylethene, **70**, 47

1,1-Diphenylethylene, **70**, 179

Disodium dihydro-1,2,4,5-tetrazine-3,6-dicarboxylate, **70**, 80

cis-, trans-2,6-Disubstituted piperidines, **70**, 57

2,6-Dithiaadamantane, **70**, 126

1,3-Dithiolane, 2-styryl-; **70**, 240

n-Dodecylbenzenesulfonyl azide, **70**, 97

Enzyme mimics, **70**, 136

Ethane, 1,2-bis(2-iodoethoxy)-; (36839-55-1), **70**, 132

1,2-Ethanediol; (540-63-6), **70**, 240

1,2-Ethanediol, 1,2-diphenyl-, [R-(R*, R*)]-; (52340-78-0), **70**, 47

Ethanol, 2-(2-aminoethoxy)-; (929-06-6), **70**, 129

Ethyl acetoacetate, **70**, 94

Ethyl bromoacetate, **70**, 248

Ethyl diazoacetate, **70**, 80

Ethyl diazoacetoacetate, **70**, 94

Ethyl 5-(N,N'-diethylaminomethyl)-3,4-diethylpyrrole-2-carboxylate, **70**, 74

Ethyl 3,4-diethylpyrrole-2-carboxylate, **70**, 70

Ethylene, 1,1-diphenyl-, **70**, 179

Ethyl formate, **70**, 232

Ethylidene transfer reagent, **70**, 187

Ethyl isocyanoacetate, **70**, 70

ETHYL 2 METHYL-5-PHENYL-3-FURANCARBOXYLATE, **70**, 93

ETHYL 4,4,4-TRIFLUOROTETROLATE, **70**, 246

Ethyl 4,4,4-trifluoro-2-(triphenylphosphoranylidene)acetoacetate, **70**, 246

Ferric chloride, **70**, 180

Flash chromatography, **70**, 4

Formaldehyde; (50-00-0), **70**, 71, 112

Formic acid, azodi-, diethyl ester, **70**, 10

Formic acid, chloro-, benzyl ester; (501-53-1), **70**, 29

Formic acid, chloro-, trichloromethyl ester; (503-38-8), **70**, 72

Formic acid, cyano-, methyl ester, **70**, 256

Formic acid, ethyl ester; (109-94-4), **70**, 232

Formic acid, oxydi-, di-tert-butyl ester, **70**, 18, 140, 152

Furan, **70**, 97

3-Furancarboxylic acid, 2-methyl-5-phenyl-, ethyl ester; (29113-64-2), **70**, 93

Furan, tetrahydro-, compd. with borane (1:1); (14044-65-6), **70**, 131

3-Furoic acid, 2-methyl-5-phenyl-, ethyl ester, **70**, 93

(-)-Galantic acid, **70**, 27

Galloles, **70**, 276

Germoles, **70**, 276

Glutaraldehyde, **70**, 54

Glycerol, **70**, 140

Glycine; (56-40-6), **70**, 30

Gooch tube, **70**, 226

Halo lactams, **70**, 106

Heterocyclic azadiene, **70**, 85

Heterodienophiles, **70**, 86

3-Hexanol, 3-nitro-; (5342-71-2), **70**, 68

3-Hexanol, 4-nitro-, acetate; (3750-83-2), **70**, 69

4-Hexyn-1-ol; (928-93-8), **70**, 111

4-Hexyn-1-ol, methanesulfonate; (68275-05-8), **70**, 111

4-Hexyn-1-yl methanesulfonate, **70**, 111

1-Hexyne; (693-02-7), **70**, 216

1-HEXYNYL(PHENYL)IODONIUM TOSYLATE, **70**, 215

Hydantoin, 1,3-dibromo-5,5-dimethyl-, **70**, 151

Hydrogen cyanide, **70**, 56

Hydrogen peroxide; (7722-84-1), **70**, 265

N-[2-(2-Hydroxyethoxy)ethyl]acetamide, **70**, 130

threo-β-Hydroxy-L-glutamic acid, **70**, 27

Hydroxy lactams, **70**, 106

1-Hydroxy-2-methylpent-1-en-3-one, **70**, 232

(E)-2-(4-HYDROXY-6-PHENYL-5-HEXENYL)-1H-ISOINDOLE-1,3(2H)-DIONE, **70**, 195

2,4-Imidazolidinedione, 1,3-dibromo-5,5-dimethyl-; (77-48-5), **70**, 151

Imines, hydrolysis of, **70**, 36

Indolizidines, **70**, 57

8-exo-IODO-2-AZABICYCLO[3.3.0]OCTAN-3-ONE, **70**, 101

IODOLACTAMIZATION, **70**, 101

Iodomethane: Methane, iodo-; (74-88-4), **70**, 19, 113, 178

2-(3-Iodopropyl)-1H-isoindole-1,3(2H)-dione, **70**, 195

Iodosobenzene, **70**, 217

Iron, tetracarbonylbis (η^5-2,4-cyclopentadien-1-yl)di-, (Fe-Fe); (38117-54-3), **70**, 177

1H-Isoindole-1,3(2H)-dione, 2-(3-bromopropyl)-; (5460-29-7), **70**, 195

1H-Isoindole-1,3(2H)-dione, 2-(4-hydroxy-6-phenyl-5-hexenyl)- ; (121883-31-6), **70**, 195

1H-Isoindole-1,3(2H)-dione, 2-(3-iodopropyl)-; (5457-29-4), **70**, 195

Isoprene; 1,3-Butadiene, 2-methyl-; (78-79-5), **70**, 256

Isopropylmagnesium chloride; (1068-55-9), **70**, 157

2(1H)-Isoquinolinecarboxylic acid, 3,4-dihydro-6,7-dimethoxy-1-methylene, 1,1-dimethylethyl ester; (82044-08-4)**70**,139

Isoquinoline, 3,4-dihydro-6,7-dimethoxy-1-methyl-; (4721-98-6), **70**, 139

ISOQUINOLINE ENAMIDES, **70**, 139

Ketene silyl acetals, **70**, 162

Keto carbenoids, **70**, 97

β-Ketoesters, C-acylation, **70**, 256

Kieselgel, **70**, 13

Kieselgel 60 silica gel, Merck, **70**, 258

β-Lactone, **70**, 1

Lead tetraacetate, for oxidative ring expansion of isoquinoline enamides, **70**, 139

Lewis acids catalysts, **70**, 65

Liquid ammonia, **70**, 256

2-Lithiofuran, **70**, 154

N-Lithiopyrrole, **70**, 154

2-Lithiothiophene, **70**, 154

Lithium; (7439-93-2), **70**, 205

Lithium aluminum hydride, **70**, 55

Lithium chloride; (7447-41-8), **70**, 196

Lithium, methyl-; (917-54-4), **70**, 165

Magnesium, chloro[(trimethylsilyl)methyl]-; (13170-43-9), **70**, 241

Methane, nitro-; (75-52-5), **70**, 183, 184, 226

Methanesulfonic acid, trifluoro-, trimethylsilyl ester; (27607-77-8), **70**, 101, 234

Methanesulfonyl azide, **70**, 97

Methanesulfonyl chloride; (124-63-0), **70**, 111

L-Methionine; (63-68-3), **70**, 31

L-Methionine, methyl ester, hydrochloride; (2491-18-1), **70**, 29

L-Methionine, (N-[(phenylmethoxy)carbonyl]-, methyl ester; (56762-93-7), **70**, 29

9-Methoxy-9-borabicyclo[3.3.1]nonane, **70**, 173

1-Methoxy-2-methylpenten-3-one, **70**, 237

(E,Z)-1-METHOXY-2-METHYL-3-(TRIMETHYLSILOXY)-1,3-PENTADIENE, **70**, 231

(S)-(-)-α-Methylbenzylamine, **70**, 35, 38

Methyl L-2-(benzyloxycarbonylamino)-4-(methylsulfinyl)butanoate, **70**, 30

Methyl cyanoformate, **70**, 256

2-Methylcyclohexanone, **70**, 35

3-Methyl-2-cyclohexen-1-one, **70**, 261

2-METHYL-1,3-CYCLOPENTANEDIONE, **70**, 226

Methylene transfer reagent, **70**, 187

Methyllithium-lithium bromide complex, **70**, 165

4-Methylmorpholine N-oxide, **70**, 47

(R)-(-)-10-METHYL-1-OCTAL-2-ONE, **70**, 35

(R)-(+)-2-Methyl-2-(3-oxobutyl)cyclohexanone; (91306-30-8), **70**, 36

METHYL (1α,4aβ,8aα)-2-OXODECAHYDRO-1-NAPHTHOATE, **70**, 256

Methyl 3-phenyl-5-carboxamidopyrrole-2-carboxylate, **70**, 85

2-Methylpiperidine, **70**, 265

6-METHYL-2,3,4,5-TETRAHYDROPYRIDINE N-OXIDE, **70**, 265

Methyl vinyl ketone, **70**, 35

Michael-type alkylation, **70**, 40

Mitsunobu conditions, **70**, 6

Molybdophosphoric acid, **70**, 31

Morpholine, 4-methyl-, 4-oxide; (7529-22-8), **70**, 47

Mosher ester, **70**, 24, 48

p-Naphthalenesulfonyl azide, **70**, 97

2(3H)-Naphthalenone, 4,4a,5,6,7,8-hexahydro-; (1196-55-0), **70**, 256

2(3H)-NAPHTHALENONE, 4,4a,5,6,7,8-HEXAHYDRO-4a-METHYL-,
    (R)-; (63975-59-7), **70**, 35

Nickel-catalyzed coupling reactions, **70**, 243

Nickel, dichlorobis(triphenylphosphine)-; (14264-16-5), **70**, 241

1-Nitrocyclohexene, **70**, 77

4-Nitro-3-hexanol, **70**, 68

Nitromethane, **70**, 183, 184, 226

NITRONES, **70**, 265

1-Nitropropane, **70**, 69

Nitrosomethylurea, **70**, 22

2,3,7,8,12,13,17,18-OCTAETHYLPORPHYRIN, **70**, 68

Octahydrobinaphthol, **70**, 65

$\Delta^{1(9)}$-Octalone-2, **70**, 256

ORGANOBIS(CUPRATES), **70**, 204

Osmium tetroxide; (20816-12-0), **70**, 47

Oxalyl chloride; (79-37-8), **70**, 101, 121, 205

Oxazolidine, **70**, 57

3,4-Oxazolidinedicarboxylic acid, 2,2-dimethyl-, 3-(1,1-dimethylethyl) 4-methyl ester, (S)-; (108149-60-6), **70**, 19

3-Oxazolidinecarboxylic acid, 4-formyl-2,2-dimethyl-, 1,1-dimethylethyl ester, (S)- or (R)-; (S)- (102308-32-7); (R)- (95715-87-0), **70**, 18

5H-Oxazolo[3,2-a]pyridine-5-carbonitrile, hexahydro-3-phenyl-, [3R-(3α,5β,8aβ)]-; (88056-92-2), **70**, 54

2-Oxetanone, 3-amino-, (S)-, 4-methylbenzenesulfonate; (112839-95-9), **70**, 10

OXIDATION OF SECONDARY AMINES, **70**, 265

Pentanedial; (111-30-8), **70**, 54

Pentanoic acid, [1,1'-binaphthalene]-2,2'-diyl ester, (±)-; (100465-51-8), **70**, 61

3-Pentanone; (96-22-0), **70**, 232

Pentanoyl chloride; (638-29-9), **70**, 61

1-Penten-3-one, 1-hydroxy-2-methyl-; (50421-81-3), **70**, 232

4-Pentyn-1-ol; (5390-04-5), **70**, 113

Perfluoroalkyl acetylenes, **70**, 246

1,10-Phenanthroline; (66-71-7), **70**, 206, 274

Phenylacetylene; (536-74-3), **70**, 94, 157, 218

(E)-5-PHENYLDODEC-5-EN-7-YNE, **70**, 215

2-(2-Phenylethenyl)-1,3-dithiolane, **70**, 240

(-)-Phenylglycine, **70**, 55

(-)-Phenylglycinol; (56613-80-0), **70**, 54

2-(Phenylsulfonyl)tetrahydro-2H-pyran, **70**, 157

1-PHENYL-2,3,4,5-TETRAMETHYLPHOSPHOLE, **70**, 272

1-Phenyl-1-(trimethylsiloxy)ethylene, **70**, 82

Phosphine, triphenyl-; (603-35-0), **70**, 1, 248

1H-Phosphole, 2,3,4,5-tetramethyl-1-phenyl-; (112549-07-2), **70**, 272

Phosphomolybdic acid, **70**, 22

Phosphonium, (carboxymethyl)triphenyl-, bromide, ethyl ester, **70**, 246

Phosphonous dichloride, phenyl-; (644-97-3), **70**, 273

Phthalimide, N-(3-bromopropyl)-, **70**, 195

Piperidine, 2-methyl-; (109-05-7), **70**, 265

Pirkle Type 1-A column, **70**, 64

Polyaza-crown compounds, N-alkyl-substituted, **70**, 135

Polyaza-crowns, **70**, 135, 136

Poly(triphenylmethyl)methacrylate on silica gel (Chiralpak OT), **70**, 64

Polyvalent iodine compounds, **70**, 222

Porphine, 2,3,7,8,12,13,17,18-octaethyl-; (2683-82-1), **70**, 68

Porphyrins, synthetic, **70**, 74

Potassium cyanide; (151-50-8), **70**, 55

Potassium iodide-starch test paper, **70**, 267

Potassium permanganate, **70**, 55

Propane, 1-nitro-; (108-03-2), **70**, 69

1,2,3-Propanetricarboxylic acid, 2-hydroxy-; (77-92-9), **70**, 54

1,2,3-Propanetriol; (56-81-5), **70**, 140

Propionaldehyde, **70**, 69

Propionitrile, 2,2'-azobis[2-methyl-; (78-67-1), **70**, 151

Propionyl chloride, **70**, 226

2H-Pyran, tetrahydro-2-(4-pentynyloxy)-; (62992-46-5), **70**, 113

2H-Pyran, tetrahydro-2-(phenylethynyl)-; (70141-82-1), **70**, 157

2H-Pyran, tetrahydro-2-(phenylsulfonyl)-; (96754-03-9), **70**, 157

Pyrazole, **70**, 1, 2

1H-Pyrazole; (288-13-1), **70**, 1

1H-Pyrazole-1-propanoic acid α-[[(phenylmethoxy)carbonyl]amino]-, **70**, 1

3,6-Pyridazinedicarboxylic acid, 4-phenyl-, dimethyl ester; (2166-27-0), **70**, 82

Pyridine, 2,3,4,5-tetrahydro-6-methyl-, 1-oxide; (55386-67-9), **70**, 265

Pyrimido[1,2-a]azepine, 2,3,4,6,7,8,9,10-octahydro-; (6674-22-2), **70**, 70

Pyrrole; (109-97-7), **70**, 151

1H-Pyrrole-1-carboxylic acid, 2-bromo-, 1,1-dimethylethyl ester;
    (117657-37-1), **70**, 151

1H-Pyrrole-2-carboxylic acid, 3,4-diethyl-, ethyl ester; (97336-41-9), **70**, 70

1H-Pyrrole-1-carboxylic acid, 2-(trimethylsilyl)-, 1,1-dimethylethyl ester;
    (75400-57-6), **70**, 151

1H-Pyrrole-2,5-dicarboxylic acid, 3-phenyl-, dimethyl ester; (92144-12-2), **70**, 82

Pyrrole, 3,4-diethyl-; (16200-52-5), **70**, 68

Rhodium(II) acetate dimer, **70**, 93

Robinson-Schopf condensation, **70**, 56

Selenophenes, **70**, 276

Serine, **70**, 6

Serine, N-carboxy-, β-lactone, benzyl ester, L-; (26054-60-4), **70**, 1

Serine, N-carboxy-, N-tert-butyl ester, L- , **70**, 11, 19

L-Serine, N[(1,1-dimethylethoxy)carbonyl]-; (3262-72-4), **70**, 11, 19

L-Serine, N-[(phenylmethoxy)carbonyl]-; (1145-80-8), **70**, 1

Serine: L-Serine; (56-45-1), **70**, 18

Silane, (chloromethyl)trimethyl-; (2344-80-1), **70**, 241

Silane chlorotrimethyl-; (75-77-4), **70**, 152, 164, 216

Silane, 1-hexynyltrimethyl-; (3844-94-8), **70**, 215

Silane, [[1-(2-methoxy-1-methylethenyl)-1-propenyl]oxy]trimethyl-, (Z,E)-; (72486-93-2), **70**, 231

Silane, tetrachloro-; (10026-04-7), **70**, 164

Silane, trimethyl(4-phenyl-1,3-butadienyl)-, (E,E)-; (70960-88-2), **70**, 240

Silane, trimethyl[(1-phenylethenyl)oxy]-; (13735-81-4), **70**, 82

Silicon chloride, **70**, 164

Siloles, **70**, 276

Siloxy dienes, **70**, 237

Silyl enol ethers, **70**, 162

SILYLOLEFINATION, **70**, 240

Simmons-Smith reaction, **70**, 186

Sodium; (7440-23-5), **70**, 177

Sodium amide, **70**, 256

Sodium azide; (26628-22-8), **70**, 94

Sodium-benzophenone ketyl, **70**, 23

Sodium hydride; (7646-69-7), **70**, 231

Sodium iodide; (7681-82-5), **70**, 112, 195

Sodium metabisulfite, **70**, 48

Sodium periodate, **70**, 30

Sodium taurocholate, **70**, 61

Sodium tetrafluoroborate, **70**, 178

Sodium tungstate dihydrate, **70**, 265

D-erythro-Sphingosine, **70**, 27

Spiroacetals, **70**, 160

SPIROANNELATION, **70**, 204

Spiro[4.5]decan-7-one, 9,9-dimethyl-; (63858-64-0), **70**, 204

Stannoles, **70**, 276

STEREOSPECIFIC COUPLING WITH VINYLCOPPER REAGENTS, **70**, 215

Stilbene, (E)- **70**, 47

Stiboles, **70**, 276

2-Substituted dihydropyrans, **70**, 160

2-Substituted tetrahydrofurans, **70**, 160

Succinic acid, **70**, 226

Sulfide, chloromethyl methyl; (2373-51-5), **70**, 177

Sulfonyl azide, **70**, 93

Sulfonyl chloride, **70**, 93

Sulfur chloride; (10545-99-0), **70**, 121

Swern oxidation, in presence of sulfur moiety, **70**, 126

Tetrahydrobenzazepine, **70**, 145

TETRAHYDRO-3-BENZAZEPIN-2-ONES, **70**, 139

Tetrahydro-2-(4-pentynyloxy)-2H-pyran, **70**, 113

TETRAHYDRO-2-(PHENYLETHYNYL)-2H-PYRAN, **70**, 157

Tetrahydropyranyl ethers, **70**, 160

1,2,4,5-Tetrazine-3,6-dicarboxylic acid, 1,2-dihydro-; (3787-09-5), **70**, 80

1,2,4,5-Tetrazine-3,6-dicarboxylic acid, 1,2-dihydro-, dimethyl ester; (3787-10-8), **70**, 80

1,2,4,5-Tetrazine-3,6-dicarboxylic acid, 1,2-dihydro-, disodium salt; (96898-32-7), **70**, 80

s-Tetrazine-3,6-dicarboxylic acid, dimethyl ester; (2166-14-5), **70**, 79

2-Thiaadamantane, **70**, 126

(endo,endo)-9-Thiabicyclo[3.3.1]nonane-2,6-diol, **70**, 121

9-Thiabicyclo[3.3.1]nonane-2,6-diol, (endo,endo)-; (22333-35-3), **70**, 121

9-THIABICYCLO[3.3.1]NONANE-2,6-DIONE; (37918-35-7), **70**, 120

2-Thiabrexane, **70**, 126

Thiacycloheptane, **70**, 126

Thiacyclohexane, **70**, 126

Thionyl chloride; (7719-09-7), **70**, 80, 130

Thiophenes, **70**, 276

Thymine polyoxin, **70**, 27

p-Toluenesulfonic acid, **70**, 71

p-Toluenesulfonic acid monohydrate; (6192-52-5), **70**, 13, 19, 217, 233

Toluenesulfonyl azide, **70**, 97

Transition metal carbene complexes, **70**, 186

Triaza-21-crown-7, **70**, 132

Tributylstannane, **70**, 166

Trichloromethyl chloroformate, **70**, 72

Trifluoroacetic acid; (76-05-1), **70**, 11, 141

Trifluoroacetic anhydride, **70**, 246

(E,E)-TRIMETHYL(4-PHENYL-1,3-BUTADIENYL)SILANE, **70**, 240

1-Trimethylsilyl-1-hexyne, **70**, 215

(Trimethylsilyl)methylmagnesium chloride, **70**, 241

Trimethylsilyl trifluoromethanesulfonate, **70**, 101, 234

Triphenylphosphine, **70**, 1, 248

Triphenylphosphine: Phosphine, triphenyl-; (603-35-0), **70**, 10

Trisilane, 1,1,1,3,3,3-hexamethyl-2-(trimethylsilyl)-; (1873-77-4), **70**, 164

TRIS(TRIMETHYLSILYL)SILANE, **70**, 164

Urea, N-methyl-N-nitroso-; (684-93-5), **70**, 22

L-Vinylglycine hydrochloride, **70**, 30

Zinc; (7440-66-6), **70**, 195

Zinc bromide; (7699-45-8), **70**, 55, 157

Zinc-copper couple, **70**, 186

Zinc dust, **70**, 85

Zirconium, dichloro-π-cyclopentadienyl-, **70**, 272

Zirconium metallacycle, **70**, 274

Zirconocene dichloride, **70**, 272

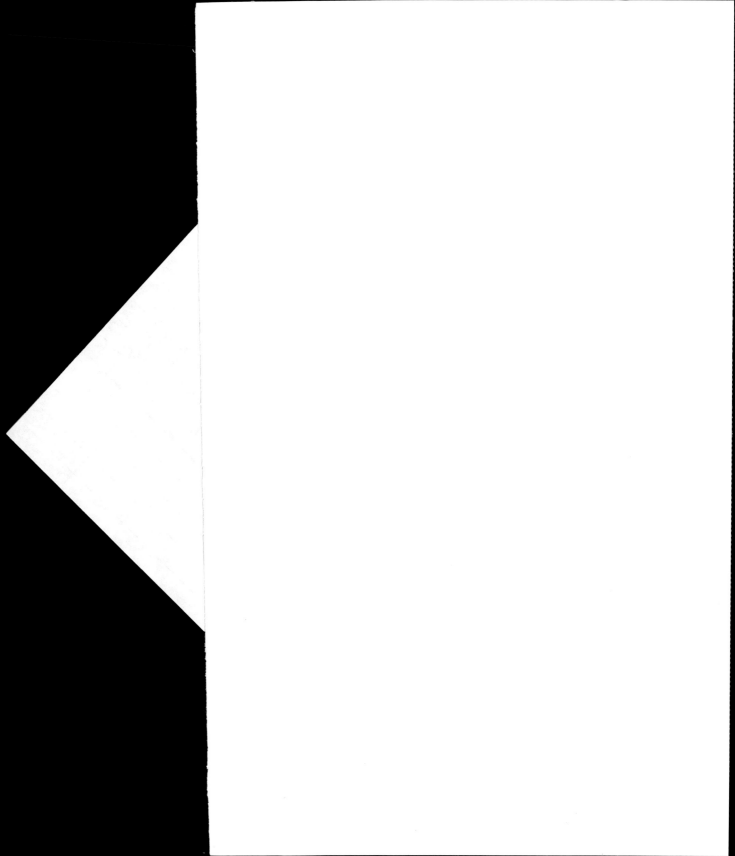

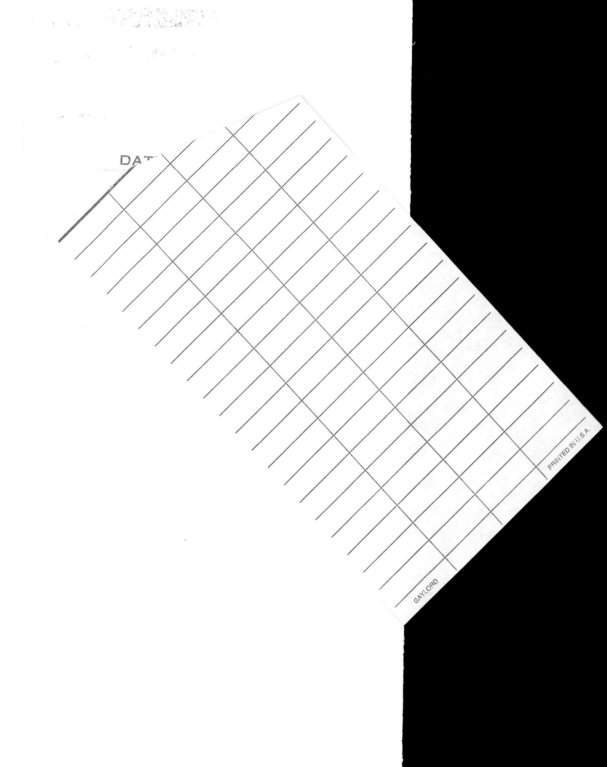